✣ SANTA BARBARA'S FLYING A STUDIO ✣

Santa Barbara's

Flying A Studio

Stephen Lawton

FITHIAN PRESS
SANTA BARBARA ♣ 1997

Photographs on pages 9, 18, 46, 48, 50, 51, 52, 53, 54, 77, 79, 98, 110, and 127 courtesy of the Santa Barbara Historical Society.

Photographs on pages 33, 49, and 155 courtesy of the author.

Documents on pages 41, 47, and 71 courtesy of the University of California, Santa Barbara, Library.

Document on page 37 courtesy of the University of Southern California Library.

Copyright © 1997 by Stephen Lawton
All rights reserved
Printed in the United States of America

Published by Fithian Press
A division of Daniel and Daniel, Publishers, Inc.
Post Office Box 1525
Santa Barbara, CA 93102

Book design: Eric Larson

LIBRARY OF CONGRESS CATALOGING-IN-PUBLICATION DATA
Lawton, Stephen, (date)
    Santa Barbara's Flying A studio / by Stephen Lawton
      p.   cm.
    Included bibliographical references.
    ISBN 1-56474-210-5 (alk. paper)
    1. American Film Manufacturing Company—History. 2. Motion picture industry—California—Santa Barbara—History.   I. Title
PN1999.A5L39 1997
384'.806579491—dc21   96-39089
                                                          CIP

Acknowledgements

The story of Flying A Studio began because there seemed to be knowledge of this long-ago time in Santa Barbara, but no real understanding of how it fits into the larger picture of the era of silent films. With the help of the Department of History at California State University at Fresno, the Fresno State University Library, the UCSB research library, and the Santa Barbara Historical Society, I was on to a fascinating journey for information to place into my knowledge of motion picture history. When the work was finally completed several years ago, my family and friends encouraged me to get this story published. My eternal gratitude to my wife, Jackie, for her endless words of encouragement. Jackie and her sister Mary Watté provided the editing for this first-time writer. My thanks for showing me the road to a good transition and how an active voice is best in whatever I say. My gratitude to a good friend and fellow author, Ron McGriff, for getting me to just do it and send the manuscript to a publisher. And finally heartfelt thanks and appreciation to those wonderful individuals who consented to share their memories with me. They will always be remembered for their willingness to relive those long-ago days.

Contents

Introduction ✤ 11

Background—Studios Move to the West Coast ✤ 15

The Flying A Arrives in Santa Barbara ✤ 31

Santa Barbara's Theaters and Motion Pictures ✤ 61

Santa Barbara and the Flying A ✤ 69

Conclusion ✤ 93

Appendix A: Memories of the American Film Company ✤ 95

Bibliography ✤ 123

The "Flying A" logo

Introduction

THE AMERICAN FILM Manufacturing Company of Chicago established its western branch in Santa Barbara, California, in July 1912. This book explores the reasons behind this move and shows how it had a reciprocal impact on the company and its host town.

Historical film scholarship is relatively new. Although books on film appeared as early as the late 1920s, they were few, and were the exception rather than the rule. It is only since the 1960s that magazine articles and books have presented a history of the first years of the motion picture industry. These studies provide lessons in historical causation, as historians deal with questions of how different cinematographic techniques affected the industry, how society influenced the movies in content and style, and how financial issues affected the growth or decline of various studios.

Nevertheless, documentation and research of the American Film Manufacturing Company—or Flying A, as it is nicknamed after its logo—are still lacking. Books on film history give only a line or two on Flying A. Santa Barbara history books devote only one or two paragraphs to it. Articles in magazines are not only very scarce, but are also superficial in content. Even the most extensive research work on this studio, Timothy's Lyons' *The Silent Partner*, a doctoral dissertation of the early 1970s, limits its investigation to the internal workings of the company and is not concerned with the company's role in the community.

One of the most important primary sources of information about Flying A and its activities is the Santa Barbara *Morning Press*, one of two daily newspapers of that era. In this book, I include much information from the *Morning Press* because it reflects the local viewpoint, which focuses not so much on the American Film Company's finances, but on how it affected the community. The impact of the American Film Company on Santa Barbara parallels the way film became part of the American lifestyle in cities across the country.

Many of the events described in the *Morning Press* substantiate recollections by participants of extensive filming in areas near Santa Barbara. Press reports document filming in the Santa Ynez Valley, Lompoc, at the Gibraltar Dam, and at numerous other nearby locations.

Favorable articles concerning the company in the *Morning Press* helped American Film gain a positive reception in Santa Barbara. It should be noted, however, that the studio undoubtedly encouraged favorable reports in the newspaper by supplying information to the *Morning Press* itself.

The initial success of the American Film Company in Santa Barbara stimulated inquiries from other interested film companies. Although the studio was unsuccessful in the long run, news stories prove that the industry looked upon Santa Barbara as a suitable location for not one, but many film companies. Santa Barbara welcomed the new industry, and news reports in the *Morning Press* announced the formation of the local Santa Barbara Motion Picture Company with the same enthusiasm with which they had welcomed the American Film Company.

This book documents changes within the theater pages of the *Morning Press*, showing the impact of the motion picture industry on the press. Since newspaper sales influence the editor's choice of what to print, the space devoted to the motion picture industry reflects the community's acceptance and appreciation of this new form of art. The columns and regular news articles provide a direct connection with the growth of motion pictures in Santa Barbara. The theater pages established the "'Flying A' Notes" column for regular announcements of American Film news. Similar treatment was given the Santa Barbara Motion Picture Company in the newspaper's "Ess Bee Notes."

The theater section of the newspaper also reflected the growth and decline of five highly competitive theaters in Santa Barbara. Motion picture advertisements show which theaters presented the American Film Company's motion pictures to the audiences. With an understanding of how the exchange organizations bought films from the stu-

Introduction

dios and rented them to the theaters on a contract basis, a study of the competition between theaters can be presented just from researching advertisements in the *Morning Press*.

Finally, personal interviews conducted with residents of Santa Barbara who remember the days of the American Film Company are collected in an appendix. Memories of personal involvement in the American Film Company's productions, the personnel of the studio, and how well the community received the film company support the accounts in the *Morning Press*.

This book shows the emergence of a new industry—the motion picture—and its impact on Santa Barbara. The *Morning Press* documents the growth and fall of the American Film Company, the interest of other film companies in Santa Barbara, and the change in theater programs. Santa Barbara provides a fascinating study of changes occurring in the the early years of the motion picture industry.

✤ CHAPTER 1 ✤

Background— Studios Move to the West Coast

BEFORE BEGINNING the study of the American Film Manufacturing Company,[1] a summary of the reasons for the motion picture industry's move to the West Coast is necessary to shed light on the reasons for one company settling in Santa Barbara. In addition, an observation of the struggles within the industry will give some understanding of the patent wars and the origins of independent film companies. Methods of film distribution will be discussed because the American Film Manufacturing Company and other independent companies grew out of the film exchanges.

Films of the 1890s depicted scenes of famous events, such as President McKinley's inauguration, or views of national wonders, such as Niagara Falls. As narrative film replaced the early topical scenes so common at the turn of the century, studios hurried to meet the public's increasing demand for exciting fictional stories on the screen.

Many studios were founded in these early years to meet the demand for the short subject. These films of one or two reels (approximately twelve and twenty-four minutes in length) extended the dime novel of the late 1890s to the film format. Using a very simple story line, they provided plenty of action for the limited length of the film. Innovative film techniques used by such directors as Edwin S. Porter and D.W. Griffith developed the narrative film into a powerful tool for the storyteller. With the cross-cut, the flashback, and fast-paced editing, the director kept his audiences coming back to the local nickelodeon.

Many film companies were established on the East Coast and in Chicago as businessmen realized the profits to be made in the new industry. These early studios were Edison (founded in 1893), American Mutoscope and Biograph (1896), Lupin (1897), and Vitagraph (1897). The number of film companies gradually increased after the turn of the century to include Kalem in 1907 (named after the owners, Kleine, Long, and Marion), William Selig's Polyscope company, and Spoor and Anderson's Essanay Film Manufacturing Company (also named after the owners). The East Coast and Chicago served as early centers of the industry.[2]

The first decade of the century witnessed rapid growth in the infant industry as the new motion picture medium gained acceptance by audiences. Motion picture entertainment created fortunes for the opportunist. In the scramble to get to the top, rapid success or failure was possible. Competition within the industry resulted in conflict in many areas. Prior to World War I, film companies waged business wars over patents, distribution rights, and access to exhibitors.

The dispute over patents occurred between Edison and the Biograph company. They both claimed the rights to film cameras and projectors. This dispute was resolved in the courts with the formation of Edison's trust of companies (The Motion Picture Patents Company, September 1908) given exclusive rights to the equipment. Anyone not in the trust who used the equipment was liable to prosecution. So this Motion Picture Patents Company effectively controlled the legal production of all motion pictures. Many "outlaw" film studios continued to operate, however, in defiance of the trust. In response, the trust tried to control the film distribution system.[3]

At that time, motion pictures were distributed by the film exchange, the middleman between the studios and the theaters. The exchange purchased the films from the studios, providing a market for the studios, and then turned around and rented the product to the theaters. The Miles brothers of San Francisco came up with this new idea. They would purchase films from the studios and rent a package of films of varying quality at a lower price to the exhibitors on a state's rights system. The exchange distributors had a territory (usually a state) in which they were licensed to distribute films to exhibitors. By 1907, between 125 and 150 exchanges existed.[4] This was to the theaters' advantage because they no longer had to purchase the expensive films from the studios.

After gaining control of cameras, projectors, and Eastman film stock, Edison's trust attempted to take over the film exchanges in order to get the "outlaw" studios in line. In turn, the film exchanges broke with the trust and began to distribute films from these independent studios.

The two men most successful in this struggle against the trust were Carl Laemmle and William Fox. Timothy Lyons, in *The Silent Partner*, lists the sources of the independent films as six companies from the United States and five companies from Europe.[5]

Edison trust exchanges tried to obtain films from independent filmmakers as well as from the trust studios. Two unlicensed exchanges were the Western Film Exchange, owned by John Freuler and Harry Aitken, and the H. and H. Film Exchange, founded by Samuel Hutchinson and Charles Hite. Later, these four men played key roles in the founding of the American Film Company, one of the early film studios to arrive in California. American Film, based originally in Chicago, transferred its studios to Santa Barbara after short stays in Colorado, Arizona, and other locations in California.[6]

The Edison trust eventually created the General Film Company to purchase the unlicensed film exchanges and control them. This led the United States government to pursue legal action against the trust, but other factors were also at work. The independent film companies developed strength of their own. Lyons states that independents benefited from the trust-busting spirit of the country. They usually localized operations with the exchanges rather than with the mammoth trust, used the new "star system" (attracting noted actors to the screen with high salaries and featured billing), and were flexible in pricing their films.[7]

In 1909, the independents formed their own organization, the National Independent Moving Picture Alliance (IMP), led by Carl Laemmle. A year later the IMP reorganized by having the New York Motion Picture Company of Adam Kessel and Charles Baumann join the Western Film Exchange of Freuler and Aitken. This new film organization became the Motion Picture Distributing and Sales Company. Now many of the independents thought they saw two trusts instead of one.[8]

Samuel S. Hutchinson, 1917.

Later in 1909, a third major film exchange organization appeared. The Associated Independent Film Manufacturers included the film studios of Thanhouser, Centaur, and others. After all this realignment, the two independent factions compromised and joined under the name Motion Picture Distributing and Sales Company late in 1910.⁹

Studios Move to the West Coast

In 1912 the independents quarreled again, and the Motion Picture Distributing and Sales Company split into two parts. One group of independents went with the Mutual Film Corporation, headed by Harry Aitken, and other independents joined the Universal Film Company, under the control of Carl Laemmle.[10]

During this period the film exchanges began to develop their own film companies. The exchanges realized it was more profitable to make their own films than to purchase motion pictures from established film companies. In 1910 Edwin Thanhouser and Charles Hite established Thanhouser Films, Samuel Hutchinson and John Freuler formed the American Film Manufacturing Company, and Harry Aitken created his Majestic Film Company.[11]

California had been active in the distribution and exhibition businesses before the arrival of the eastern studios. One of the first theaters to show films exclusively was the Cineograph Theater on San Francisco's Market Street, which showed its first film in 1897. This predates the 1902 shift to films by Tally's Theater in Los Angeles. The Miles brothers of San Francisco were first to open a film exchange. Instead of theaters buying films from the studios, the Miles brothers offered to buy the studios' films and rent them out to the theaters. They added to their inventory of other studios' films by making some films of their own. They filmed scenes of San Francisco and environs such as those shown in A *Trip Down* Mt. *Tamalpais*, a documentary-type film depicting a train going down a winding railroad track. A camera was mounted on the train, giving the audience a magnificent view of the countryside.[12]

Lyons describes this period of early film history as a businessman's game. Constant struggling over control of distribution and exhibition disrupted the industry. Simultaneously, the audience demand for film increased throughout the period of struggle among the film corporations. Films went to two-reelers, the star system gained acceptance, and the production of feature films was initiated. But especially important was the movie industry's first tentative move toward the West Coast.

Perhaps the thesis that Los Angeles was a safe refuge from patent forces began with the landmark history of motion pictures, *A Million and One Nights*, a book written in 1926 by Terry Ramsaye. Ramsaye believes that independent film studios moved to California to avoid harassment from the patent forces. "Cameras vanished from under the noses of the guards. Mysterious chemical accidents happened in laboratories, resulting in the loss of negatives." So, Ramsaye states, in 1907 the Selig company sent Francis Boggs to Los Angeles to begin filming on the West Coast.[13]

Later historians agree that the escape of independents from patent forces was a major element in the move west. Anthony Slide states that after the conclusion of the patent wars with the court decision of 1908, the independents escaped to the West Coast, beyond the reach of the marshals representing the Edison trust, who pursued them as illegal users of Edison cameras.[14]

Historians also suggest that Los Angeles was ideal specifically because of its proximity to Mexico. In his autobiography, Adolph Zukor agrees that escape from the patent marshals was the reason for the independents' move to the West Coast. Zukor refers specifically to Mexico: "History is not absolutely clear on the point, but it appears that Southern California was first 'discovered' for movies by small producers who wanted to be able to skip across the Mexican border at a moment's notice."[15]

Even in Kevin Starr's recent history of California, Los Angeles is portrayed as having been a haven for the independents. Starr explains why the movie industry left San Francisco for the Los Angeles area, using the remarks of screenwriter Anita Loos. Loos believes that San Francisco provided the movie industry with theatrical and literary talent, along with excellent weather. But Los Angeles provided something else: its proximity to Mexico. San Francisco was a "legal" town, easily accessible from the East Coast, whose law officers joined readily with Edison lawyers and marshals to search for illegal equipment.[16]

The proximity of Los Angeles to the Mexican border may have been a factor in the independents' move to that city, but simple geography refutes the likelihood of its being a major reason. The Mexican border is approximately 150 miles from the Los Angeles area, making it an impractical escape route from patent marshals.

Lillian Gish has a different perspective on this controversy: "In those days when the producers were fighting the patent forces, no one had to run to Mexico. You must remember there were not even forty-eight states, much less fifty."[17] She refers to the fact that Arizona and New Mexico did not enter the Union until 1912. Or perhaps the isolation of the small towns of Southern California was enough to bring comfort to the independents' desire for privacy.

Historian Gerald Mast speculates that besides scenery and distance from Edison and the marshals, the incentive for the move to Los Angeles may have included a favorable real estate market. Mast believes that early studios settled in the suburbs of Los Angeles (Hollywood, Culver City, and Burbank) because real estate developers sold property to the studios at "ridiculously low prices." Thousands of acres were purchased for the offices, buildings, processing plants, and horse and cattle ranches necessary for a major studio to function. Developers realized how the new industry could benefit the Los Angeles area in terms of growth and employment.[18]

Another explanation for the movement to the southern West Coast is lower labor costs. The Los Angeles Chamber of Commerce, under publisher Harrison G. Otis's leadership, waged an anti-union campaign to keep wages below those in San Francisco and on the East Coast. Union organizers and pickets went to jail on conspiracy charges. This confrontational atmosphere climaxed with the dynamiting of the *Los Angeles Times* building in October 1910. Prosecution and conviction of the unionist bombers served to prevent unionization for nearly three decades. The success of Los Angeles businessmen in blocking the growth of unions probably appealed to prospective motion picture companies.[19]

The first attempt to film in the Los Angeles area took place in 1906. George Van Guysling and Otis M. Grove opened their studio and filmed their outdoor scenes on a ranch outside the Los Angeles area in what later became Hollywood. But apparently this was a short-lived company.[20] Another historian credits the Horsely brothers from New Jersey with being the first to film in Hollywood. Unfortunately, no date is available for their film entitled *The Law of the Range*, filmed at Sunset Boulevard and Gower Street.[21]

The Selig organization filmed *The Count of Monte Cristo* in Los Angeles in 1907. On Olive and 7th streets, the Selig company made the first dramatic motion picture filmed entirely in California, entitled *The Power of the Sultan*. The company built a studio in Los Angeles at 3800 Mission Road, in Eastlake Park, and later at 1845 Alessandro Street (now Glendale Boulevard). Adam Kessel and Charles Baumann established a studio in 1907 at 1712 Alessandro Street, which later became the site of the Keystone studios of Mack Sennett. The Biograph Company, with D.W. Griffith, placed its West Coast studio on Pico and Georgia Streets in Los Angeles.[22]

Studios were located in areas surrounding the Los Angeles community, in nearby towns and villages selected for their scenic beauty or for enticements offered by real estate developers. The first studio in Hollywood was the Nestor company, the West Coast branch of the Centaur Film Company, which set up a place of operation on Sunset Boulevard and Gower Street in 1911. The following year, Nestor was taken over by Universal Studios. The New York Motion Picture Company established its studio in 1909 in Edendale (now Glendale or Echo Park). Thomas Ince established his studio in Santa Monica in 1908 (Inceville) and filmed short William S. Hart westerns. Vitagraph arrived in Santa Monica in 1911.[23]

Further information on the film industry's move to California comes from the autobiography of Mrs. D.W. Griffith (Linda Arvidson). After citing the first studios to arrive in California, she describes Biograph's move to the West Coast. She states that Selig's Polyscope Company arrived in Los Angeles in early 1908 (*sic*) under the leadership of Frank Boggs, and that the New York Motion Picture Patents Corporation had a unit called the Bison Company in Los Angeles in 1909 under the direction of Fred Balshofer. The Kalem Company was also established in California, although Arvidson does not give the date of its arrival.[24]

Arvidson's husband, D.W. Griffith, joined the Biograph Company as director during its winter seasons. Biograph sent a unit for a short stay in California during the winter season of 1909–1910; it arrived after Christmas in 1909. The studio was set up on the corner of Grand Avenue and Washington Street, and remained for about three months. Arvidson reflects, "After shivering through one Eastern winter, trying to get the necessary outdoor scenes for our pictures, we concluded that it would be to our advantage to...hie ourselves to the city of Los Angeles."[25] She offers an interesting description of the Los Angeles area at the time of their arrival. Hollywood "didn't mean a thing.... [Pasadena was where] millionaires sojourned for two months during the Eastern winter."[26] There was a mission in San Fernando, but "it was rather far away, and right in the heart of a parched and cactus-covered desert."[27]

After Griffith moved his studio to Los Angeles, Lillian Gish joined him in February 1913. She remembers that, "In New York...Griffith had often found the weather a handicap to production. Moreover, he needed a variety of backgrounds for his films, which eliminated New York as a year-round headquarters."[28] The Griffith studios were at the corner of 12th Street and Georgia Avenue in Los Angeles. Gish believes that many of the westerns were filmed in Chatsworth Park.[29]

In her autobiography, Gish states that Griffith spent three winter seasons filming in California before her arrival in Los Angeles with the company in February 1913 for the fourth winter. Gish recalls: "Los Angeles, then a residential city of about 300,000 people, had wide boulevards, churches and Spanish-style houses. Nearby were ocean and desert, snow-capped mountains and green valleys, Spanish missions and fruit farms."[30]

Beaches, mountains, and deserts offered many opportunities for film studios. The use of Spanish-style homes and missions opened new themes for the film story. Besides the San Fernando mission, the Biograph unit used the San Juan Capistrano and San Gabriel missions for locations on Spanish theme pictures. *Ramona* was filmed in the town of Camulos in Ventura County.[31]

Adolph Zukor settled on Hollywood as the location for his studios, on the corner of Sunset and Hollywood boulevards. He states in his autobiography that "there was no particular reason [for the choice of location]. It was an undeveloped suburb of Los Angeles, mostly orange and lemon groves. The chief attraction was a rentable farmhouse suitable for dressing rooms, a small laboratory, and offices."[32] Zukor states that Cecil DeMille and his Jesse Lasky Feature Play company (later to become Paramount) filmed *The Squaw Man* in Hollywood beginning on December 29, 1913. Zukor writes that his two early Mary Pickford films were already being filmed in Hollywood by that time, but he was unable to give exact dates for these.[33]

Southern California was not the only area on the West Coast where early studios were established. Anderson's Essanay Company filmed along the Central Coast as he searched for new locations for his westerns. The American Film Company followed him up the California coast and settled in Santa Barbara.

The Essanay studio of Chicago, which began in 1907 under the leadership of George K. Spoor and G.M. "Broncho Billy" Anderson, was determined to make westerns in a real western location. Anderson filmed in Golden, Colorado, in 1908 and later moved to the Los Angeles area, then to Santa Barbara, and finally to Niles (near Oakland), California. There he made many of his famous Broncho Billy westerns.

In a press release to a trade journal, Essanay explained its rationale for traveling throughout the West in search of locations:

> We do not believe it is possible to faithfully produce Western pictures—call them cowboy picture if you will—without going right out into the districts which the film is supposed to portray. The people of the United States, in particular, insist on Western films being really Western films, and that the Essanay Company has been successful in their efforts is proved by the fact that their Western films are accepted and approved by Western audiences, who would be the last people in the world to accept substitutions artificially staged in the East.[34]

Other factors played a role in Essanay's settlement near San Francisco. First, Anderson wanted an area where outdoor filming could take place year-round. Second, San Francisco's and Oakland's opera houses offered an abundance of theatrical talent to draw upon for extras in the films. And third, direct access to the express rail lines to Chicago and the East was possible in the Bay Area to speed the transport of negatives to the processors in Chicago. The processing had to be accomplished as fast as possible in the highly competitive business of motion pictures. Later, when film laboratories were built on the studio sites, rapid transportation to the East lost importance.[35]

Many featured players at Essanay became famous actors in later years, working in studios in the Los Angeles area. For a time, the Bay Area attracted the notable names of the industry. Actors employed by Essanay included Francis X. Bushman and Henry B. Walthall (famous for his starring role in *Birth of a Nation*). In addition, the famous French comedian Max Lindner made a few films for the studio. Charlie Chaplin even had a one-year contract in 1915. He made a series of fourteen short films for the Essanay studios.[36]

In 1911 Essanay began producing films in Marin County and in 1913 constructed new studio buildings as production expanded. The prime years of its operation were 1912–1916.

After this peak period, however, Essanay closed its doors because of the distribution problems, Anderson's extravagance, and competition from other western film cowboys. By 1916, Essanay had encountered financial difficulties because no distributing organization could be found which permitted them even a reasonable profit. Essanay lost when competing with other companies that controlled distribution as well as production. By that time, Anderson had adopted a freewheeling, carousing lifestyle in San Francisco and was spending large amounts of money. And finally, the westerns of Tom Mix and William S. Hart offered stiff competition for the Broncho Billy films.[37]

The Anderson Film crew from Essanay spearheaded the search for sunshine and western locations. Their travels to California and up the Pacific Coast brought other motion picture companies to the shores of California. The Melies film company in Paris (Star Film) sent Gaston Melies to California in 1911. He followed the Broncho Billy Essanay film company to Santa Barbara but rejected the site as too expensive for his crew and not exactly the scenery that he wanted for his films.

Melies found the community of Santa Paula more to his liking. Beginning in April, he made several one- and two-reel shorts using the Southern Pacific Railroad and various locations in the Santa Paula area. The following year Melies built his second small studio in Santa Paula and produced more films until the property was sold in August 1913.

In 1912, the International Photoplay Company built an outdoor stage and stock corrals in hopes of establishing a major studio in Santa Paula. But by May of the following year, the firm was out of business.

The St. Louis Motion Picture Company purchased the Melies property and began production in September 1913. After filming a series of westerns, released through Universal, the St. Louis Motion Picture Company sold its studio and property to Willis Robards. He established the Robards Film Manufacturing Company, which produced only a few films before its demise a few months later.[38]

The American Film Manufacturing Company arrived in Santa Barbara just a few miles to the north in July 1912. This Chicago organization also followed "Broncho Billy" Anderson's film unit up the coast. For approximately the next two years, it was American Film (the Flying A studio) that provided the Broncho Billy westerns with their strongest competition.

In 1901, a literary journal, *The Overland Monthly*, expressed the appeal of Santa Barbara to any interested readers. In the article describing Santa Barbara, the journal shared the "fact that on this bit of southern coast, under these sheltering mountains, by this sapphire sea, is found a charm of ocean and sky, of mountain, island and shore, of tropical growth and temperate airs, such as no other nook of this old planet possesses—a charm which holds one in a thrall of happy enjoyment that lasts the whole year round."[39]

The Kalem, Essanay, and American Film studios exemplified the move by filmmakers to locations chosen specifically for their scenery. Kalem's film crews searched for beautiful locations in Ireland, Rome, and the Middle East. This was an unusual practice for the makers of the short fiction films at that time. Essanay and American Film moved from Chicago to Colorado and on to California in search of authentic locations for their westerns. The public needed to be satisfied. Seeing the real West on the screen appealed to the audience of those early years.

Considering the experimental filmmaking in San Francisco, the work of Essanay in the Niles Canyon area, and the early work of the American Film Manufacturing Company (Flying A) along the Central Coast of California, it seems strange that the studios did not remain in those areas, but chose instead to relocate to Los Angeles.

Part of the reason may lie in San Francisco's 1906 earthquake and fire. Convincing new companies to establish studios in the area after the quake was a difficult task. Many studios resisted building in San Francisco proper. In 1915, the trade journal *Moving Picture World* described the earthquake as being "responsible for keeping San Francisco off the film manufacturing map for a number of years, perhaps depriving the coast city of eminence in the industry.[40] They may have been willing to establish their studios out in Marin County or Oakland, as was Anderson; but generally, for people from the East, moving to Los Angeles was a less threatening prospect than locating in the Bay Area. Besides, San Franciscans were focusing their energies on rebuilding their city, not on establishing film studios.

One independent company which did survive in the Bay Area was the California Motion Picture Corporation (CMPC), under the ownership of Nevada silver baron Herbert Payne. Beginning in 1912 with advertising films, CMPC moved into the production of feature films. This studio preferred the Marin County area because of the scenic locations. Even though Southern California attempted to bring in the industry through the use of tax breaks and utility benefits, the Bay Area was able to hold onto some studios.

The CMPC had a ranch near San Rafael, used for indoor scenes, and another ranch near Boulder Creek, used for exterior shots. The Boulder Creek location is where one of the first western streets was built. This studio began a newsreel entitled the *Golden Gate Weekly*, which was shown in vaudeville houses in the Bay Area. The newsreels' subject matter varied greatly and included depictions of hunting along the Sacramento River and the latest productions at the Greek Theater in Berkeley. One newsreel was especially interesting, a documentary of the life of the California Indian Ishi, who had recently been discovered in the wilds of northern California. Only still photographs of this film exist today. The CMPC's first feature film was *Pageant of San Francisco*, in 1914. In this and subsequent productions, the company chose California themes for its motion pictures.[41]

Each studio moving to the West Coast faced the realistic challenge of finding the best locations, obtaining distribution for their films, and remaining a financially successful enterprise. In time, even the beautiful scenery in Santa Barbara and San Francisco could not solely sustain the independents. These studios had led the move to the West Coast in search of locations, sunshine, and isolation from patent marshals. Eventually, as major studios consolidated and were able to distribute their own films, theaters accepted fewer of the independents' motion pictures, which had to be handled through the distributing companies. As actors and directors joined the move to Southern California looking for career advancement in the expanding industry, and problems of distribution worsened, the number of Northern Californian independent film companies declined and failed.

NOTES

1. The company is referred to variously, by different sources, as the American Film Company, American Film, the American Film Manufacturing Company, and Flying A.
2. Timothy J. Lyons, *The Silent Partner: The History of the American Film Manufacturing Company 1910–1921* (New York: Arno Press, 1974), 19.
3. Lyons, 22–23.
4. Lyons, 20.
5. Lyons, 25–26.
6. Lyons, 25–26, 69.
7. Lyons, 29–30.
8. Ibid.
9. Lyons, 31.
10. Lyons, 31–32.
11. Lyons, 33–34.
12. Bell, 100–103.
13. Terry Ramsaye, *A Million and One Nights: A History of the Motion Picture Through 1925* (New York: Simon & Schuster, 1926; reprint, New York: Touchstone Books, 1986), 532–34.
14. Slide, *Early American Cinema*, 81.
15. Adolph Zukor and Dale Kramer, *The Public Is Never Wrong: The Autobiography of Adolph Zukor* (New York: G.P. Putnam's Sons, 1953), 55.
16. Kevin Starr, *Inventing the Dream: California Through the Progressive Era* (New York: Oxford University Press, 1975), 288–89.
17. Lillian Gish, letter to the author, 5 March 1988.
18. Gerald Mast, *A Short History of the Movies*, 4th ed. (New York: Macmillan, 1986), 98.
19. Kevin Brownlow and John Kobal, *Hollywood: The Pioneers* (New York: Alfred A. Knopf, 1979), 91. For an excellent article on unions in the early days of motion pictures, see Michael C. Nielsen, "Labor Power and Organization in the Early U.S. Motion Picture Industry," *Film History: An International Journal* 2 (June–July 1988): 121–32.
20. Starr, 124.
21. Ralph J. Roske, *Everyman's Eden: A History of California* (New York: Macmillan, 1968), 494.
22. Joseph H. North, *The Early Development of the Motion Picture (1889–1909)* (New York: Arno Press, 1973), 264. See also, Brownlow and Kobal, 90, 97; Kevin Brownlow, *The Parade's Gone By...* (New York:

Alfred A. Knopf, 1968; reprint, Los Angeles: University of California Press, 1975), 30–32.
23. Brownlow and Kobal, 90–91, 165; Smith, 3.
24. Mrs. D.W. Griffith (Linda Arvidson), *When the Movies Were Young* (1925; reprint ed., New York: Benjamin Blom, 1968), 143.
25. Mrs. Griffith, 143.
26. Mrs. Griffith, 146.
27. Ibid.
28. Lillian Gish and Ann Pinchot, *The Movies, Mrs. Griffith, and Me* (1969; reprint ed., Englewood Cliffs, N.J.: Prentice–Hall, 1975), 81.
29. Gish and Pinchot, 81–82.
30. Gish and Pinchot, 82.
31. Zukor and Kramer, 155; also see pp. 161–67 for an excellent chapter on work at the missions; pp. 169–70 for a description of *Ramona* film.
32. Zukor and Kramer, 101.
33. Zukor and Kramer, 101–102, 122.
34. "In the Far West," *The Bioscope*, 9 February 1911; quoted in Slide, *Early American Cinema*, 66–67.
35. Geoffrey Bell, *The Golden Gate and the Silver Screen* (New York: Cornwall Books, 1984), 42–43.
36. See Bell, 65–80, for a description of Essanay.
37. Bell, 58–69. Also see Donald Packhurst, "Broncho Billy and Niles, California: A Romance of the Early Movies," *The Pacific Historian* 26 (Winter 1982): 1–22, for an excellent article on Essanay in Niles.
38. See Wallace E. Smith, "Santa Paula's Film Days," *The Ventura County Historical Society Quarterly* 16 (Winter 1971), 1–24, for an overview of Santa Paula's role in early film history.
39. C.M. Gidney, "About Santa Barbara County," *The Overland Monthly* 38 (August 1901): 164.
40. *Moving Picture World*, 10 July 1915, 248; quoted in Bell, 104.
41. Bell, 70–74, 77.

✤ CHAPTER 2 ✤

The Flying A Arrives in Santa Barbara

SANTA BARBARA became a center for motion picture production for a very short time when the American Film Manufacturing Company (Flying A) arrived there on July 6, 1912.

The development of the American Film Manufacturing Company in Santa Barbara reflected the changes in the industry that took place throughout the United States. The American Film Manufacturing Company evolved from a one- and two-reel motion picture studio to one that produced occasional features and serials. In a competitive bid for "stars" in motion pictures, the studio featured players such as Jack Kerrigan and Mary Miles Minter in its productions.

Besides hosting the American Film studio, Santa Barbara itself exemplified the evolution of theaters from stage and vaudeville entertainment to the inclusion of moving pictures in their programs. In fact, moving pictures became the primary focus of many of the Santa Barbara theaters in the highly competitive atmosphere of that period.

The history of moving pictures in Santa Barbara can be discovered in the pages of its primary newspaper, the *Morning Press*. References to employment opportunities, a payroll to be spent within the city, and growth of the community's businesses prevail in early articles about the new film studio.

In addition, the *Morning Press* left clues about the changes within the theaters of Santa Barbara reflecting growth and popularity of motion pictures. Advertisements found in the *Morning Press* depict the evolution of moving pictures from supplemental programs in vaudeville houses to the establishment of theaters devoted exclusively to moving pictures. In the decade between 1910 and 1920, Santa Barbara reflected the change occurring throughout the country as the moving picture business grew into a major industry.

The small Santa Barbara community offered beautiful scenery and excellent weather to the motion picture entrepreneur. A 1901 issue of *The Overland Monthly* described the city:

> Its location at the foot of the highest peaks of the Santa Ynez range, with an outlook toward the southwest [sic] and the beautiful bay that is said by travelers to be almost a counterpart of that of Naples, Italy, leaves nothing to be desired. South of the city and forming a protection against the fresher breezes from the Channel, lies the "Mesa," a range of hills from 300 to 400 feet in height. Between this mesa and the foothills of the Santa Ynez, on an inclined plane with an average slope of about 100 feet to the mile, the rose-embowered homes of 7,000 people are found, scattered over an area of about 3,000 acres. Thus many of these homes have very ample grounds, and nowhere is the population crowded. With rare judgement the Franciscan friars who founded the Missions of California selected this locality as the site of their most important post.[1]

Santa Barbara was a growing community before the American Film Company arrived. Based on the 1910 census of 11,659, the *Morning Press* estimated the population to be approximately 15,000 in 1913.[2]

Santa Barbara had an attractive climate for an industry needing adequate lighting. In 1912, interior lighting for films was in its infancy. The number of days of sunshine became important to the prospective filmmaker. *The Overland Monthly* observed that in Santa Barbara, on the average, only one day in six could be termed "cloudy." In 1900, "the mean temperature of the three winter months was 57.4 deg., that of the spring months 58.3 deg., of the summer months 64.8, and the fall months 64.1 deg."[3] According to the article, Santa Barbara experienced "on the average an annual rainfall of but seventeen inches, distributed over eight months of the year."[4] One might come there "to escape the heats of summer as [well] as the frosts of winter."[5]

By the turn of the century, Santa Barbara was attracting a great many visitors because of its climate and scenery. As a resort community, the town offered the visitor excellent accommodations. The prestigious Arlington Hotel, built in 1875 and remodeled in 1911, used the latest in modern conveniences. The Potter Hotel catered to the

The Arlington Hotel.

tourist who desired a view of the beach. Many other boarding houses and hotels offered visitors rooms and meals for their stay in Santa Barbara. One such boarding house, Edgerly Court, on the corner of Sola and Chapala Streets, seemed to enjoy success because of the film industry; it opened one year after the arrival of the American Film Company and rented rooms to several of the Flying A personnel.

During the first ten years of the twentieth century, Santa Barbara's transportation and communication facilities expanded to link it with other towns up and down the coast. The community had recently gained rail connections with San Francisco and Los Angeles. Previously, the short Pacific Coast Railway extended from the small town of

Los Olivos to San Luis Obispo. Access to Los Olivos had been only by stagecoach northward over the San Marcos Pass until the turn of the century. But by the days of the Flying A studio, automobiles crossed the San Marcos Pass to Los Olivos in four hours. An extension of the Southern Pacific Railroad along the coast and up to San Luis Obispo provided uninterrupted rail service from Los Angeles to San Francisco and eliminated the need to travel over the mountains and into the Santa Ynez Valley to get to Los Olivos and rail service northward.

Within the city, an electric rail system offered service to both visitors and citizens of Santa Barbara. Recent gas and electric plants provided the latest power technology to the homes and businesses of the community.

Santa Barbara was an active town: it promoted sports, entertainment, culture, and education. The popular sports of the day, golf and polo, attracted community enthusiasts. Local clubs provided ample grounds for both sports. Various fraternal organizations had been formed in Santa Barbara by the early 1900s. A Union Club, the Santa Barbara Club, the Country Club, and a women's club actively participated in the community. Many different lodges and religious organizations were thriving. There were schools from the elementary level up through high school.

City news was reported in two daily newspapers, the *Morning Press* and the *Independent,* and two weekly papers, one of which was the *Morning Weekly.* Santa Barbara's papers reflected a knowledgeable and thriving community.[6]

The movie industry experienced such a tremendous evolution of its product and such an overwhelming popular acclaim nationwide that the existence of live theater was soon being threatened. Santa Barbara, like other cities of the day, experienced the motion picture phenomenon at the turn of the century with the advent of films in their theaters. At first, the three or four theaters established in the city by 1910 presented short films along with vaudeville acts, as evidenced by advertisements for entertainment in the *Morning Press.*[7]

Moving pictures caused some concern among the people in plays. The theaters showing the motion pictures charged a lower admission than those presenting live stage plays. On June 9, 1911, the Associated Press carried a story out of Chicago entitled "War Is Waged on Moving Pictures." Balcony and gallery prices were "slashed" at the

The Flying A Arrives in Santa Barbara

downtown theaters to ensure attendance of the "gallery gods." Charles Frohman, noted theatrical impresario, stated, "Very many people go to the moving picture shows simply because they cannot afford to see plays." Frohman commented further that the last time he talked to the stage actor Henry Irving, Irving said he missed seeing his "gallery gods," the common people who were the "best barometer of public opinion."[8]

The newness of the industry was not allowed to wear off because directors pleased audiences time and time again with shifts in subject matter and the innovations of filming on location. During 1900–1907, the western had become so popular with audiences that it replaced the travelogue films of previous years. D.W. Griffith directed several early western shorts for Biograph, as did other directors in studios such as Kalem. But the Broncho Billy westerns at Essanay of Chicago surpassed the competition from the American Film Manufacturing Company, also of Chicago. Now all studios sought real western locations. Essanay filmed Broncho Billy westerns at scenic "out-west" locations. Traveling first to Colorado and Arizona, Essanay eventually ended up in the Niles Canyon area near Oakland, California. "Broncho Billy" Anderson did shoot some films in the Santa Barbara area in 1910; he arrived by train with a boxcar equipped with everything necessary for filming. But Anderson continued on up the California coast to his future site of the Essanay studio, Niles.[9]

The American Film Company of Chicago had established three film genres in the first year of its operation: westerns, social comedies, and dramas.[10]

The American Film Company units making westerns were isolated from the Chicago headquarters as they searched for the "real" western locations in Colorado, Arizona, and California. When director Allan Dwan received a cable to join the American Film Company's western unit, he remembers that "the new organization had a problem. Somewhere in California—no one knew where—was one of their companies. The supply of films had dried up and, despite frequent cables, so had the supply of information."[11] The western units of the eastern studios were on their own in those days.

Searching for locations for westerns, the studio had first sent a unit to Colorado. Not content with that location, the unit moved on to Arizona and finally to California. They filmed in San Juan Capistrano,

Lakeside, and La Mesa (both of the latter are near San Diego). After exhausting these locations, a unit arrived in Santa Barbara during July of 1912. The choice was primarily because of the locations available for shooting a variety of films. By 1913, American Film was producing two-reel films. The future for the studio looked bright as it built new facilities in Santa Barbara.[12]

The attitude of the studio toward finding scenic locations for its westerns was captured in a press release issued by American Film in 1911:

> There is no artificial lighting system in a studio that can approach the perfection of sunlight. There is no scenic artist who has ever been born who could originate and paint scenery that will anywhere equal the stupendous beauties of the Western country. We propose to install into Flying A Cowboy films the spirit of the West as it really was, and the romance of the Western life as everybody, even the Westerners themselves, fondly imagine it had ought to be.[13]

One person who certainly appreciated Santa Barbara's sunshine and scenery was Robert Phelan, a cameraman for Flying A during the 1914–1915 period. His most notable film for the studio was *Purity*, with Audrey Munson (famous for posing nude for statuary displayed at the 1915 Panama–Pacific Exposition in San Francisco) as the "revealing" star. In an interview, Mrs. Phelan stated that the American Film Company arrived in Santa Barbara "looking for a variety of scenery."[14]

The seriousness with which the studio sought the best locations for their westerns is apparent from the statements by Roy Overbaugh, a cameraman for the Flying A studio:

> In 1910 they sent their first production unit to the Coast, and I was in charge of photography. We located in La Mesa, a few miles outside of San Diego. I might say here that at this early date artificial light had not been developed to the point where it was suitable for proper lighting. Consequently, all scenes were photographed by daylight. This meant that scripts and scenarios had to be written so that no interior scenes were necessary. Later on, outside stages were devised consisting of a

Page from Moving Picture World with announcement of Flying A westerns.

crude platform with an overhead covering of diffusing cloth such as nainsook. This made interior sets possible, but even then they were seldom used unless there was no alternative and daylight was still the necessary source of light. This, of course, plus certain scenic advantages, was the principle [sic] reason why the early motion picture companies chose California. It was a matter of business: they had to locate where they could depend upon the greatest number of sunny days. Well, to continue, after working out of La Mesa for about two years and photographing exterior scenes exclusively, we practically ran out of settings. We had photographed nearly all the worthwhile scenery within a reasonable working radius.... The result...was the selection of Santa Barbara as the best possible base of operations, considering climate, sunshine and variety of scenery.[15]

Wallace Kerrigan, the business manager of American Film, visited Santa Barbara in June of 1912 after receiving an enthusiastic report from Allan Dwan. Kerrigan negotiated the lease of an ostrich farm for a base of operations. This acreage served as a headquarters and a corral for stock used in the westerns. The Flying A Company felt it had exhausted the locations around the small community of La Mesa twelve miles north of San Diego. Kerrigan reported to the *Morning Press* that the intention was to film in Santa Barbara as an experiment, but possibly to locate there permanently in the future. Since the exposure laboratory was in La Mesa, the negatives would be shipped to La Mesa for development. When the La Mesa laboratory lease expired, it might be possible to move the entire operation to Santa Barbara. Later newspaper articles and interviews reported that the studio shipped the negatives to Chicago for processing, a variation from this first newspaper account.[16]

As Flying A settled in Santa Barbara, newspaper accounts reflected the complexity of emotions that preceded the town's acceptance of the new and daring industry. Sometimes the press hinted at the reluctance of the civic leaders to welcome the crew and cast. Sometimes the press disclosed some moral issues the citizens raised about the actors and their lifestyles. However, the overwhelming message is one that favors the financial impact of the film industry on Santa Barbara and portrayed the

The Flying A Arrives in Santa Barbara

newcomers as decent, congenial, and hard-working entrepreneurs.

The press reported some interesting comments Kerrigan made in an interview on June 18, 1912:

> I can say for the company that we have a nice lot of people. When we first came to La Mesa the preacher of a church warned his congregation against the moving picture people. Now everybody is trying to induce us to remain.[17]

The Flying A brought two automobiles and the equipment needed to film two screenplays a week. In essence:

> The coming of the company would mean an additional payroll in Santa Barbara. Nothing is sent out but the manufactured films, while every cent of salaries comes from the outside...the company would continually be in need of labor and supplies.[18]

Kerrigan and the owners finalized a one-year lease for the ostrich farm property. A couple of weeks later, the Santa Barbara *Morning Press* revealed that the American Film Company would locate in the city and announced their plans to build an indoor studio and darkroom.[19]

Roy Overbaugh, chief cameraman for the studio, described the building of the darkroom in an interview in 1954:

> After we had been here a few months and were still at the old ostrich farm, it was decided that time would be saved and results better if we could arrange to do our developing here instead of sending the negatives to Chicago. A vacant building on Cota Street near State was rented. Developing, fixing and washing tanks, drying drums and other necessary equipment was installed and we started developing our exposed film immediately. This was quite an advantage, as, instead of waiting a week or two to get our film back from the east, we knew definitely within twenty-four hours what results we were getting, and whether or not any scenes had to be remade.[20]

The original company, as reported in the initial announcement by the *Morning Press*, included Overbaugh, Allan Dwan as producer and manager, Wallace Kerrigan as business manager, Jack Kerrigan as leading man, Pauline Bush as leading woman, and one C.P. Morrison as the lead cowboy and property man. There were thirty-five members in the company.[21]

One flattering paragraph in the *Morning Press* article stated:

> There is the probability that this city will become the permanent home of the "Flying A" film. Most of the work is done outdoors, and Mr. Dwan and Mr. Kerrigan have thoroughly considered the scenic possibilities and consider them about inexhaustible.[22]

The July 9 *Morning Press* described the acting company as "comfortably located, pleased with surroundings and [confident] that they have finally struck their permanent home."[23] This was another optimistic note for the city fathers. The members of Flying A stayed at the Arlington Hotel during these first weeks in the city. Jack Kerrigan, who within the next two years became a notable screen idol, arrived along with his twin brother, Wallace, his sister, Kathleen, and his mother, Mrs. J.W. Kerrigan. The newspaper in 1912 announced that Mrs. Kerrigan would be "permanently located here with her two sons,"[24] which added respectability to Jack Kerrigan's standing in the community.

To calm the apprehensions of the local citizenry, who might have wondered what kind of people this acting crowd included, the *Morning Press* interviewed one member of the American Film organization:

> We are a very congenial lot…but you have no idea how tired people can get of each other. At La Mesa we would hardly see any one else but members of the company and it got on the nerves of all. It will be different here and we feel things will move smoothly. From this it must not be understood there has been any friction—that is something that would not be tolerated. After the folks of Santa Barbara get better acquainted with us they will know the company is made up entirely of ladies and gentlemen.[25]

The Santa Barbara Chamber of Commerce even received the Flying A

FILM COMPANY COME TOMORROW TO LOCATE IN SANTA BARBARA

COMPANY OF THIRTY FROM LA MESA SECURES LEASE ON FORMER OSTRICH FARM.

The members of the American Film Manufacturing company, which produces the "Flying A" film, will arrive to locate in Santa Barbara tomorrow evening, and in all probability the work of producing plays in Santa Barbara settings will start Monday.

The company is at present at La Mesa, where it has been producing plays the last year. The company decided upon a change of scene and Santa Barbara was chosen. W. W. Kerrigan, the business manager was here two weeks ago, and then concluded negotiations with the owner of the old Ostrich farm on upper State street, by taking a lease for a year. This will serve as headquarters and corral for the stock. The studio for indoor work will be constructed there, and in a couple of weeks all the films exposed will be developed in new dark rooms.

The company includes about 35 persons. The cowboys left La Mesa last Saturday with the stock and can be expected at any time, but the latest by tomorrow.

The principals and heads of departments will leave La Mesa this morning, but will pass the night in Los Angeles, leaving there in the morning for Santa Barbara.

The following are the principal persons connected with the company: Allan Dwan, producer and manager; Jack Kerrigan, leading man; Miss Pauline Bush, leading woman; George Pierlot, character; Miss Louise Lester, character; Miss Jessalyman Van Trump, ingenue; Steve Nellen, juvenile; W. W. Kerrigan, business manager; Roy Overbaugh, camera operator; R. D. Armstrong, assistant camera operator; C. P. Morrison, lead cowboy and property man; "Slim" Wilson, assistant property man.

It was the desire of Director Dwan to have brought some of the leading people to Santa Barbara so that some scenes might have been taken at the San Lucas bridge opening yesterday, but it was impossible to close business at La Mesa.

Under the lease it is assured that Santa Barbara will have the company next year, and there is the probability that this city will become the permanent home of the "Flying A" film. Most of the work is done outdoors, and Mr. Dwan and Mr. Kerrigan have thoroughly considered the scenic possibilities and consider them about inexhaustible. During the summer months only so-called "thrillers" will be produced.

The company is made up of a fine lot of men and women and in all probability they will become great favorites here.

Morning Press announcement of Flying A's move to Santa Barbara, July 5, 1912.

with a formal letter of welcome addressed to Allan Dwan:

> [The chamber] appreciates it as a compliment that you have chosen our city and county for what we hope will be the permanent home and scene of operations of the company.... Our valleys, canyons, sea shore and heights are hereby declared proper targets for your camera.[26]

One early book on Santa Barbara and Montecito revealed that the American Film Company's placement in Santa Barbara meant that between $500,000 and $1,000,000 were spent in the city annually. In addition, because of the high quality of the films, only actors of "exceptional character" had been cast.[27]

During the next few months, Flying A produced a series of one-reel films (approximately twelve minutes each). Roy Overbaugh explains:

> Our schedule was two a week. As they were completed we would send them to Chicago for processing and eventually we would get back the developed negative. We would then project this negative, edit it, make any necessary retakes, and again send it to Chicago. Finally, we would receive a completed print and then the picture was ready for national distribution. Through working together as a unit for a couple of years we had become so proficient that it took us only half a day to do a picture. If we hadn't completed it in time for a late lunch, we considered ourselves slow. This meant that we had at least four days free time each week, and then was when we all fell in love with Santa Barbara and began to appreciate the pleasures it offered.... Most of our time was spent at the beach. Working was just a sideline, and even that was fun.[28]

An August 1912 report in the *Morning Press* commented that Flying A was "a well bred lot that had already won high favor."[29] This article contained biographies of leading personalities in the company and the announcement that the interior studio was nearly completed. Favorable remarks appeared in regard to the company's residence in the city:

The Flying A Arrives in Santa Barbara

The company has now been in Santa Barbara a little more than a month and its members have impressed the residents here. The leading actors have become recognised [sic] as fine men and women, and the others who fill positions incidental to the taking of moving pictures have made many friends. The cowboys are now all married if the count is correct. Four have wedded since leaving La Mesa. Some of the boys were a bit lonesome at first here and longed for the southern village, but when they learned the company would not return, they did the next best thing—married the girls. There are no happier sets in Santa Barbara than these cowboys and their brides.[30]

One former assistant make-up girl with the American Film Company, Mrs. Loiz Huyck, recalls the Edgerly Court as "practically the center of the…social activities of the movie people because Edgerly Court was built at the same time they were building the Flying A. And they had a great big ballroom."[31] She remembers the Flying A personnel living:

> at the Upham [Hotel] or Edgerly Court with an apartment of their own. If they were married, they could live at Edgerly Court because they'd have someone to do their cooking. There was no restaurant attached to the Edgerly Court area. And most of the moving picture people went over to the Upham to eat or to the restaurant…. But there were kitchenettes attached to every little apartment. They were quite livable.[32]

In regard to social activities, Mrs. Huyck explains: "Every night there was a dance…most of us were pretty well supervised, and we didn't go out on [the] beach…or wahooing around in the automobiles."[33]

Working in a film company presented a stable employment for the actor.

> The players always know where they are at. There is no traveling or shifting about…. [The salaries] are far above the average of the "legitimate" theater. There are several in the Santa Barbara company that exceed $100 a week. It would appear a shame to tell just what some are getting. A little

more than $1,000 is distributed weekly among twenty people.³⁴

The local citizens of Santa Barbara appeared in several of the early films of Flying A. One such "bit" actor, Arleigh Adams, recalls his "acting" experience:

> I was supposed to be a kid who always ate…green apples and got sick in bed. Then camera, action—they started up…but the kids start clowning around…. I started grinning and I blew it. He [the director] kicked me out of the bed and looked for a "sicker" kid. That ended my career in the movies right there.³⁵

Mrs. Loiz Huyck also acted in crowd scenes. She recalls:

> I dressed beautifully and…wore long dresses with trains…. I worked as an extra on party and beach scenes…. I'd drag in a few of my friends so I'd have somebody to talk to because I wasn't going to stand there and be dumb…it was fun and it paid a little money, but not too much.³⁶

Charlotte Burton, a local actress who achieved some prominence, appeared in the early Flying A productions between 1912 and 1916. She played supporting roles in the major serials. Local Santa Barbara historian Ike Bonilla remembers Charlotte as "a very high-grade type of person…. Of the people that they used [in the films] out of Santa Barbara as actors and actresses, she was the lead one in the most films."³⁷

The most important director in these early months in Santa Barbara was Allan Dwan. He remembers this period in film directing as a time never to be repeated.

> In those days we had full control of our companies, with no interference from the producers, subproducers, supervisors, and front offices that came later. We did what we liked and we hired whoever we liked. That's how I got Marshall Neilan, Victor Fleming, and fellows like that in the business. It's unheard of today, to walk out and see somebody you like the looks of, and say, "Come on, come and work with me." He has

to do an apprenticeship, join a union, pass all kinds of muster, and do four thousand other things—even then he can't get in.[38]

Allan Dwan's story of his start at American Film's location in San Juan Capistrano testifies to the informality of operation in the early days. He arrived from the East to locate the unit and find out what happened to them because the Chicago office had not heard from the Western unit in weeks.

> [Dwan remembers,] "They had no director because he was an alcoholic. He'd gone to Los Angeles on a binge and left the company flat. So I wired: 'Suggest you disband company—you have no director.' They wired back: 'You direct.'"
>
> Faced with this sudden responsibility, Dwan called the actors together (among them was J. Warren Kerrigan) and announced: "Either I'm a director or you're out of work."
>
> Replied the actors: "You're the best damn director we ever saw."[39]

The American Film Company participated in Santa Barbara community events to foster an image of contributing to the welfare of the town. In September 1912 they helped to raise money for the local band.[40] The January New Year's celebration in 1914 was an opportunity for citizens to "see the daredevil Morrison Brothers" performing in a Flying A rodeo.[41] The Morrison brothers handled the stock of cattle and horses for Flying A. Isaac Bonilla recalls the Morrisons as "real cowboys.... They could perform and...[went] to Lompoc...[on the] Fourth of July they had a big rodeo.... Several of them participated. They were working cowboys."[42] The studio formed a baseball team as a member of the Moving Picture League, which consisted of teams from different studios in Southern California who played each other as well as local community teams. The *Morning Press* announced upcoming games such as the one between the Flying A team and the team from Universal Studios. It was a double header; the second game played between the Flying A team and the local Santa Barbara Beavers.[43] In 1916, the company opened its glass studio for public tours for the Salvation Army. An Employees' Good Fellowship Association of the Fly-

"They were real cowboys." Art Acord, Carl Morrison, Roy Sharp, and George (Pete) Morrison (center, kneeling, facing), in a western scene.

ing A sponsored this event and the dance that was held that evening on the stage of the glass studio.[44]

By November 1912 new developments had committed the Flying A studio to a more permanent residence in the city of Santa Barbara. The president of the American Film Company, Samuel S. Hutchinson, visited Santa Barbara and conferred with an architect about the feasibility of constructing permanent buildings on a new site. The company also built a developing lab for the negative prints because the exposed film sent east to be processed frequently returned as poorly developed prints. As the *Morning Press* reported, "The work at this end has been excellent, but as the operator back east may not have understood conditions under which a picture has been taken, the proper developer may not have been used at times."[45] In addition, plans were announced to establish a second unit to film non-westerns in the Santa Barbara area. The *Morning Press* reported that this meant the number

Announcement in the Morning Press, June 5, 1914, of an upcoming Flying A baseball game.

of actors needed for a second unit would bring an estimated increase of $100,000 for payroll per year. This meant more money in circulation in Santa Barbara.[46]

The November 28 issue of the *Morning Press* carried an announcement of architect J.C. Pool's commission to build the new studios. Construction was anticipated to begin in January of 1913. The newspaper announced the new construction with pride and excitement, and for good reason, since the studio was not just going to build a facility for working, but a complex that reflected the spacious mission-style architecture so typical of Santa Barbara. The studio was not trying to force itself upon the city, but knew its presence needed to be carefully integrated with the Santa Barbara community.

The American Film Company intends to make its Santa Barbara plant a model and the finest of its kind in the country and is not worrying much about the cost. Mr. Hutchinson has not limited the expense of the improvement. He informed Mr. Pool what was needed and when he gets what he wants cost will be a secondary consideration. That is the way film manufacturing companies do business these days.[47]

The new site for the permanent studios was between State and Chapala streets on West Mission Street.

An early view of the Flying A studio on Mission Street.

The plans called for an administration building, a glassed-in and curtained studio for interior scenes, dressing rooms and a green room lounge for the actors and actresses, paint, carpenter and machine shops, property rooms, art department, garage, stables and corrals for the horses, camera equipment and loading rooms, a complete laboratory for processing, together with cutting, editing and projection rooms. The grounds were beautifully landscaped and the buildings artistically arranged around a central garden and driveway. Surrounding the entire property was a high Mission wall, with the entrance on Mission Street through huge ornamental iron gates.... When this work was completed, Santa Barbara was, without any question, the home of the best equipped and most artistic studio in the country.[48]

The administration building was to have a mission tower, and a thick eight-foot wall would include the design found in walls near the Santa

A recent view of the same site.

Barbara Mission. At the same time, the "interior grounds will be a perfect dream, and every foot of building and grounds of a style that will lend themselves to the taking of moving pictures.... [It] will be as handsome as any Montecito estate."[49] Samuel Hutchinson, president of the American Film Company, even stated that he "is anxious to show Santa Barbara that he appreciates the setting and wishes to make the entire ground a credit to the company and community. In fact, this

Aerial view of Flying A studios.

comparatively new industry in Santa Barbara would be along the line of the city beautiful."[50]

By mid-1913, the *Morning Press* had established an infrequent column entitled "Notes From the 'Flying A.'" Later called "'Flying A' Notes," it provided information on the latest events concerning the studio. The first such column, on May 24, 1913, stated:

> A full crew of workmen is at present engaged on the interior of the Mission Street studio and it is expected that early next month it will be the center for all activities. All buildings have been completed with the exception of the glass studio.... The business and direction offices will be moved from the old ostrich plant, while the temporary development plant on Cota Street will be abandoned.... By July 1 it is expected everything will be in fine working order.[51]

The second and third "'Flying A' Notes" columns appeared on May 28 and 30. The May 30 column reported an incident during one studio filming in which the hero, Jack Kerrigan, was overcome by bandits and tied to a horse that galloped down the road. A dummy was substituted for the actor. "When the dummy was properly fixed [and] the horse sent galloping madly away...a few Mexican workmen saw the horse coming and were horrified—when they saw the dummy sway and strike a tree."[52] Such amusing incidents are typical of information conveyed in the "'Flying A' Notes." Later columns also discussed locations of filming or developments in the studio's business.

Jack Kerrigan

With its expansion to a new facility in early 1913, the American Film Manufacturing Company saw a bright future ahead. But in May of that year, Allan Dwan, their producer and director, left the com-

Early view of the grounds of Flying A.

pany for Universal Studios. This departure stemmed from a personality clash with Jack Kerrigan. With the new studio's construction, the plan was for two units to operate in Santa Barbara. One was to continue making westerns under the direction of Dwan, producing two one-reel westerns per week. A second unit was created to make social dramas, starring Wallace Reid. Jack Kerrigan resented the competition, as he had gained extensive popularity from his work in the earlier Dwan westerns. The studio realized this and did not want to lose Kerrigan. When Kerrigan challenged Dwan's authority, President Hutchinson decided to let Dwan go. By the summer of 1913, the studio had completed new construction, and Santa Barbara operations expanded under Dwan's replacements, Lorimar Johnston and Sydney Ayres.[53]

Ironically, Jack Kerrigan, leading man of the short western motion pictures of American, left the studio in 1913. Kerrigan received a contract with Universal, and his career remained strong for a few more years. One *Photoplay* article in 1916 referred to him as "The Great

God Kerrigan."⁵⁴ He was one of the first matinee idols of the screen. His role in the early Flying A productions (1910–1913) helped Santa Barbara's new studio succeed in its first year.

With the newly constructed facilities, American Film increased personnel and equipment and began production on a much larger scale. No longer content with making the one-reel film, the studio converted to the two- and four-reel formats. In addition, by 1913 interior lighting had improved so that an indoor studio became feasible. Klieg lights (mercury vapor tubes and carbon arcs) were available. For a time, Santa Barbara had the largest indoor studio in existence in the United States.⁵⁵

The Flying A filmed at many locations in the Santa Barbara area. Roy Overbaugh mentions the Potter and Arlington Hotels, Montecito estates such as that of William M. Graham, the Gillespie estate, and the Glendessary residence. The Santa Barbara Mission was used extensively. In addition, they used the beach and the La Cumbre and Mission Creek areas. American Film even took cameras to Santa Cruz Island, one of the islands off the Santa Barbara coast. Hope Ranch ap-

A later view (ca. 1916) of the Flying A grounds.

The glass studio is visible in the background on the right.

peared in some of the films, and other communities up and down the coast, such as Summerland, Carpinteria, and Goleta, were used by the Flying A film crew.[56]

Over the years, members of the community recalled filming in the surrounding areas. The *Morning Press* reported location filmings with specific information on the people involved, titles of films, and shooting schedules. In August of 1912, the *Morning Press* announced the release of the first Flying A production filmed locally.

> *The Greaser and the Weakling*, the first photo play made by the "Flying A" company in Santa Barbara scenery, will be released next Monday and shown throughout the United States and Europe as will regularly all others taken here.... The coming release was taken July 10 on the old Dixie Thompson ranch west of the city. The familiar buildings will be seen with occasional glimpses of the Santa Ynez range.[57]

The Flying A Arrives in Santa Barbara

Also in August 1912, a film crew accompanied the Morrison brothers and an "auto load of rough riders"[58] over San Marcos Pass into the Santa Ynez Valley to film ranching scenes of a cattle round-up. "The driving, roping, and branding of cattle is featured."[96] But the lead actor, Jack Kerrigan, had a difficult time keeping up with these cowboys. "He was handling a rifle when this was discharged accidentally and the bullet struck the ground about four inches from the [sic] foot. After that Kerrigan quit handling firearms."[60] Kerrigan and the cameraman, R.D. Armstrong, were referred to as the "tenderfeet" on the expedition. The cowboys enjoyed keeping them awake with "snake stories" late into the night.[61]

In June of 1915, the studio sent director Henry Otto with his company to Zaca Lake, north of Los Olivos in the Santa Ynez Valley. They had traveled there earlier, in April, but the weather had not cooperated. From their headquarters in Mattei's Tavern in Los Olivos, they had "motored through snow to the lake...[but the] wintry conditions" prevented adequate filming until June. The resulting film, entitled *The Zaca Lake Mystery*, was completed upon their return trip. Staying at John Liebeu's residence at the lake, the crew took "some very remarkable scenic stuff."[62] Thus, this "beauty spot of Santa Barbara County" found its way into the Flying A productions.[63]

Perhaps the beautiful scenery inspired the Flying A cameraman to be creative in his filming of many of the westerns. Noticeable in the Flying A westerns of the period were early examples of the "panning" or traveling shots. The cameraman filmed the cowboys on horseback in long extended shots from one side of the screen to the other as he followed the riders. This permitted the audience to view more of the surrounding scenery and became a filming technique common in the westerns from American Film.[64]

A few months after the arrival of Flying A in Santa Barbara, a film unit was sent to Lompoc, over fifty miles from Santa Barbara. The film, titled *The Intrusion at Lompoc*, was synopsized in the *Morning Press*:

> Lompoc, a quiet little inland town, moves peacefully along from week to week with little more excitement then [sic] the flutter of the leaves. [But] the stage coach rolls up to the one hotel...[with] a typical gambler...[who] wrought consternation

in the heart of a number of the fair sex, aroused the jealousy of the local swains and the undying hatred of a former victim.[65]

The use of so many locations around Santa Barbara County pleased citizens anticipating growth and its resulting prosperity. A few months after Flying A's arrival in Santa Barbara, the *Morning Press* reported, "Santa Barbara is now receiving a class of advertising that goes to all sections of the world."[66] Citing an article in the trade magazine *Moving Picture World*, the *Morning Press* reported that the motion picture press chronicled the "fact that the scenery is unusual" and the *Moving Picture World's* statement that "one excellent thing...[about the third Santa Barbara release]...is its views of a beautiful country, which are finely photographed."[67] In April of 1913, American Film helped publicize Santa Barbara with its announcement that "a number of original scenic photographs" would be taken to be used in advertising of the studio.[68]

The county considered using film to present an illustration of the county's attributes at the 1915 Panama–Pacific Exposition in San Francisco. Filming parts of the county became a good idea after the Flying A studio appeared in Santa Barbara. The county supervisor from Lompoc believed a motion picture of the various stages of industries throughout the county would be an impressive exhibit.

> He would show industries in their various stages...through the bean lands and walnut groves of Carpinteria and Goleta, to the fishing grounds at Surf, to the mustard fields of Lompoc and the oil wells at Santa Maria.... [Also] first class music [would] be provided while the picture [was] being run.[69]

Santa Barbara's assets as a motion picture site became well known in the film industry. Flying A's advertisement of the Santa Barbara area interested other film companies searching for scenic locations and appealing weather.

In early 1913 Dr. Schallenberg, one of the owners of Thanhouser Films, visited the American Film Company in Santa Barbara. He was considering the feasibility of establishing a studio in the community. After touring the studio and viewing locations, he said Santa Barbara would be considered as a possible site for the Thanhouser studio. The New Rochelle, New York, location lacked good scenic areas.[70] What is not mentioned in the article about Schallenberg's visit is that Thanhouser had a studio in Florida as well as in New Rochelle. The connection with American Film came about because both Thanhouser and American Film released through the Mutual Distributing Company. And above all, Charles Hite, the head of Mutual, was a principal investor in Thanhouser. But apparently, Thanhouser declined to ever set up a studio in Santa Barbara. Nevertheless, this visit by Schallenberg indicated the interest shown in the area by the moving picture industry at that time.[71]

With the interest shown by Schallenberg and the commitment by the American Film Company, Santa Barbara's appeal to the studios became clear to the industry. This community offered film directors the background for their short moving pictures. But Santa Barbara experienced another transformation of the motion picture industry with the change of programs in their local theaters. The theater pages of the *Morning Press* during the periods preceding and following American Film's arrival in Santa Barbara presented a view of this transition. Citizens of this Central Coast city witnessed not only the arrival of a studio into their community, but also a change in the type of entertainment in their theaters.

NOTES

1. C.M. Gidney, "About Santa Barbara County," *The Overland Monthly* 38 (August 1901): 167.
2. *Morning Press* (Santa Barbara), 2 April 1913, 2.
3. Gidney, 160.
4. Gidney, 163.
5. Gidney, 164.
6. Gidney, 171.
7. *Morning Press* (Santa Barbara), 20 November 1910, 3.
8. *Morning Press* (Santa Barbara), 10 June 1911, 2.
9. Walker A. Tompkins, *Santa Barbara Past and Present: An Illustrated History* (Santa Barbara: Tecolote Books, 1975), 87; Geoffrey Bell, *The Golden Gate and the Silver Screen* (New York: Cornwall Books, 1984), 41.
10. Timothy J. Lyons, *The Silent Partner: The History of the American Film Manufacturing Company 1910–1921* (New York: Arno Press, 1974), 59–64.
11. Kevin Brownlow, *The Parade's Gone By...* (New York: Alfred A. Knopf, 1968; reprint, Los Angeles: University of California Press, 1975), 97.
12. Lyons, 69.
13. Anthony Slide, *Early American Cinema* (New York: A.S. Barnes, 1970), 82.
14. Mrs. Robert Phelan, telephone interview by author, 30 March 1988. (See Appendix A for transcript.)
15. Mr. and Mrs. Roy Overbaugh, "In the Days of the Flying A," interview by W. Edwin Gledhill (*Noticias*, 17 March 1954), *Santa Barbara Historical Society* (Fall 1976): 1–2.
16. *Morning Press* (Santa Barbara), 18 June 1912, 5.
17. Ibid.
18. Ibid.
19. *Morning Press* (Santa Barbara), 5 July 1912, 3.
20. Mrs. and Mrs. Roy Overbaugh, 4.
21. *Morning Press* (Santa Barbara), 5 July 1912, 3.
22. Ibid.
23. *Morning Press* (Santa Barbara), 9 July 1912, 2.
24. Ibid.
25. Ibid.
26. *Morning Press* (Santa Barbara), 14 July 1912, 2.
27. John R. Southworth, *Santa Barbara and Montecito: Past and Present* (Santa Barbara: Orena Studios, 1920), 188.

28. Mr. and Mrs. Roy Overbaugh, 3.
29. *Morning Press* (Santa Barbara), 18 August 1912, 2.
30. Ibid.
31. Loiz Huyck, interview by author, 9 August 1987, Carpinteria, California. (See Appendix A for transcript.)
32. Ibid.
33. Ibid.
34. *Morning Press* (Santa Barbara), 18 August 1912, 2.
35. Arleigh Adams, interview by author, 16 July 1987, Santa Barbara. (See Appendix A for transcript.)
36. Huyck interview.
37. Isaac (Ike) Bonilla, interview by author, 15 July 1988, Santa Barbara. (See Appendix A for transcript.)
38. Brownlow, 98.
39. Brownlow, 97.
40. *Morning Press* (Santa Barbara), 7 September 1912, 2.
41. *Morning Press* (Santa Barbara), 1 January 1914, 2.
42. Bonilla interview.
43. *Morning Press* (Santa Barbara), 5 June 1914, 2.
44. *Morning Press* (Santa Barbara), 3 March 1916, 10.
45. *Morning Press* (Santa Barbara), 5 November 1912, 3.
46. Ibid.
47. *Morning Press* (Santa Barbara), 28 November 1912, 5.
48. Mr. and Mrs. Roy Overbaugh, 5.
49. *Moving Picture World*, 8 February 1913, 559.
50. Ibid.
51. *Morning Press* (Santa Barbara), 24 May 1913, 3.
52. *Morning Press* (Santa Barbara), 30 May 1913, 3.
53. Lyons, 75–76.
54. William Henry, "The Great God Kerrigan," *Photoplay Magazine*, February 1916, 32–36.
55. Mr. and Mrs. Roy Overbaugh, 5.
56. Mr. and Mrs. Overbaugh, 6.
57. *Morning Press* (Santa Barbara), 31 August 1912, 3.
58. *Morning Press* (Santa Barbara), 15 August 1912, 8.
59. *Morning Press* (Santa Barbara), 20 August 1912, 3.
60. Ibid.
61. Ibid.
62. *Morning Press* (Santa Barbara), 6 June 1915, 3.
63. *Morning Press* (Santa Barbara), 4 April 1915, 2.
64. Eileen Bowser, *The Transformation of the Cinema: 1907–1915*. (New York: Scribners, 1990; reprint, Los Angeles: University of California Press, 1994), 249–50.
65. *Morning Press* (Santa Barbara), 10 November 1912, 3.

66. *Morning Press* (Santa Barbara), 14 September 1912, 5.
67. *Morning Press* (Santa Barbara), 3 April 1913, 3.
68. Ibid.
69. *Morning Press* (Santa Barbara), 17 September 1912, 5.
70. *Morning Press* (Santa Barbara), 3 May 1913, 3.
71. Anthony Slide, *Aspects of American Film History Prior to 1920* (Metuchen, N.J.: Scarecrow Press, 1978), 71–73.

❖ CHAPTER 3 ❖

Santa Barbara's Theaters and Motion Pictures

WITH THE ARRIVAL of the American Film Company in 1912, Santa Barbara became actively involved in the motion picture business, and remained so for the next few years. The city had already witnessed the relatively new phenomenon of motion pictures in its theaters, as had other communities nationwide. In most parts of the country, the higher-class theaters resisted the competition from motion pictures by lowering prices of admission to keep the middle class in its "live" theaters. The lower-income and immigrant populations enjoyed motion pictures. Immigrants with new English language skills enjoyed the short films, which carried a low admission price of five or ten cents. The middle-class audience, on the other hand, was more likely to accept the new motion pictures if they were accompanied by live vaudeville. In the competition for the middle-class audience, the vaudeville theaters that also showed motion pictures used the live acts to draw in the middle-class viewer. Santa Barbara theaters followed this trend. At first, the motion pictures were shown between vaudeville acts; later, motion pictures appeared on alternate evenings in many communities. After 1910, following a precedent set earlier by Tally's Theater in Los Angeles, theaters began shifting away from vaudeville to the exclusive showing of motion pictures.[1]

The *Morning Press* advertisements showed how the programs of the theaters changed over time. In July 1908, there were four active theaters in Santa Barbara. The Potter Theater, associated with the popular Potter Hotel, brought legitimate live entertainment to Santa Barbara. Vaudeville and "legitimate" theater productions included the Potter Theater on their circuit across the country. Occasionally, speakers gave lectures or politicians gave campaign speeches there. The Potter remained a live entertainment theater until 1916, when it finally accepted motion pictures on a limited basis.[2]

Early theaters in Santa Barbara enhanced their programs of motion pictures with other forms of entertainment. The Santa Barbara Opera House in 1908 presented live entertainment with "vaudeville, Moving Pictures, [and] illustrated songs."[3] This variety appealed to a broader cross section of the public. The illustrated songs came with slides and organ accompaniment. The Unique Theater opened on July 4, 1908, for live entertainment. This theater may have been owned by D.J. Grauman of San Francisco. He opened an early theater in the Bay City in 1898 under the same name, presenting vaudeville and motion pictures. But its advertisements disappeared from the *Morning Press* after a year. Finally, there was the La Petite Theater. Advertised as "Pictorial Vaudeville," the theater presented eleven programs in one evening, accompanied "live" by organ music. The program included "illustrated songs, short comedies, and overtures—illustrated." The program changed every Wednesday and Saturday.[4]

By 1910, Santa Barbara's theaters were changing their programs and new theaters were opening. In early 1910, the Victory Theater (which in February became the Riggslee Theater with an ownership change) advertised vaudeville and motion pictures. The Potter Theater still was the place to go for live entertainment. In October the Santa Barbara Opera House featured a speech given by Hiram Johnson during his campaign for governor.[5] However, the Opera House turned to vaudeville in November.

In October of 1910, the La Petite Theater advertised its program of films in an interesting manner. The films were listed as productions of the Film de Art, Bison, and Independent Motion Pictures (IMP) studios. This indicates that the theater received its films from the independent producers following the patent decision of 1908. La Petite still presented vaudeville programs on Monday and Friday evenings and motion pictures on Wednesday, Thursday, and Saturday.[6]

Then in the month of November, there were very few advertisements for theater entertainment, particularly for vaudeville and moving pictures. The Potter Theater continued to advertise stage plays, and the Santa Barbara Opera House some vaudeville. Finally, on November 20, 1910, Tally's placed its first advertisement in the *Morning Press* as the first theater to advertise an exclusively motion picture program. On November 26 the Victory Theater followed with an all-motion-picture billing. One interesting note is that in 1902 Tally's in Los Angeles became the first theater in the United States to show only motion pictures (although historians dispute this point, citing the Miles brothers' theater, which opened November 1901 in Seattle). D.J. Grauman's 1898 Unique Theater or Walter Furst's earlier Cineograph Theater in San Francisco exhibited motion pictures with vaudeville acts.[7]

Mid-1912 advertisements for theatrical entertainment still included the Potter Theater's program of live entertainment. La Petite Theater advanced its program of "Independent Service" with four reels of motion pictures in one evening. La Petite's advertisements stated that their program had "vaudeville and moving pictures." Tally's Theater on State Street and Canon Perdido Street advertised only motion pictures.[8]

One especially interesting program appeared at Tally's in May 1912. Advertised as "How Motion Pictures Are Made and Shown," the program was described in the *Morning Press* as

> ...a peep behind the scenes, into the world of photographic magic; it will be a revelation to thousands of moving picture "fans" who are wondering how the pictures are taken, what the film book looks like, and how it is handled, and in what mysterious way it is thrown on the screen in the theater. This picture explains it all.[9]

The film was accompanied by the music of Kluge's Orchestra, and the admission prices were ten and fifteen cents.

But now there was another theater—the Mission Theater. Its *Morning Press* advertisements described "the theater different" as "The Home of Good Pictures." In 1913 the Mission's program included Selig, Essanay, and Biograph films. Advertised as the "Best of Licensed Pictures," the films apparently were not from the IMP corporation of Carl Laemmele. The films he produced and distributed were not licensed by the patent forces in the East. This trust eventually broke under the onslaught of the federal government after a suit was filed in August of 1912 and decided by the Supreme Court during the 1914 fall term. The trust worked under the title of the Moving Picture Patent Company and General Sales Company. The Biograph, Lubin, Pathe, Selig, Kalem, and Vitagraph studios were within this corporation, commonly referred to as the Edison Trust.[10]

Santa Barbara theaters changed as they accommodated the changed interests of the audience. A theater for high-class entertainment remained—the Potter. But the other theaters adapted with a mixed program of vaudeville and motion pictures. Once Tally's appeared in Santa Barbara, the "100 percent moving picture theater" was there to stay.

Some theaters underwent ownership changes, which accounted for some of the "new" theaters. On August 31, 1912, the Palace Theater opened its doors. Formerly Tally's, this theater exclusively presented motion pictures. On May 30, 1913, the La Petite reopened as the Argus. The new theater continued playing motion pictures along with vaudeville productions.[11] H. Keno Marble, who managed the Argus, received favorable comment in the *Morning Press*: "[He] has the business down from a to z.... Marble is operating the machines at the Argus and none will say that the projection of Argus pictures is not perfect."[12] In December 1913, the Mission Theater became Kuhn's Theater. The Mission program had included vaudeville acts with the moving pictures, but the Kuhn's Theater announcement of its opening program made no mention of vaudeville. Upon its reopening, Santa Barbarans viewed films for the first time on a silver screen. "Patrons observed this greatly enhanced the value of the films."[13] The program included the General Sales Service licensed films of Pathe, Vitagraph, and Edison, among others.

During this period, studios made the films and the distributors released them to the theaters. The American Film Manufacturing Company contracted with the Mutual corporation for the distribution of its films to the theaters, as did the Thanhouser, Reliance, Princess, Komic, Majestic, Kay-Bee, Broncho, Domino, Apollo, and Keystone studios.[14]

In Santa Barbara, the Palace Theater had a contract with the Mutual Film Corporation. In April 1913 advertisement for the Palace

> ### Early Santa Barbara Theaters
>
>
> Argus Theatre, 630 State Street
> California Theatre, 20 West Canon Perdido Street
> Kuhn's Theatre, 618 State Street
> La Petite Theatre, 622 State Street
> Lloyd's Theatre, 630 State Street
> Mission Theatre, 618 State Street
> Palace Theatre, 904 State Street
> Portola Theatre, 721 State Street
> Potter Theatre, 235 State Street
> Riggslee Theatre, 619 State Street
> Sta. Barbara Opera House, 33 E. Canon Perdido St.
> Strand Theatre, 630 State Street
> Tally's Theatre, 904 State Street
> Unique Theatre, 30 W. Ortega Street
> Victory Theatre, 619 State Street
>
> ---
>
> From *Noticias, vol. XXXVII no. 2, Summer 1992*

Theater lists Thanhouser, Flying A, and Keystone films and the *Gaumont Weekly* newsreel as their Mutual productions. At that point, only the Palace exclusively presented motion pictures. Any of the films available through Mutual could be seen at this theater, which was the largest Santa Barbara theater at the time.[15]

Sometimes Flying A previewed its films at the Palace. A short time after the Flying A company arrived in Santa Barbara, the cast and crew viewed the latest D.W. Griffith film, *The Resurrection*. Inspired by Leo Tolstoy's novel, the motion picture was to be shown on a state's rights basis. Wallace Kerrigan, business manager of Flying A, and Walter Griffin had secured these rights and privately screened the film for their company. This may have been one of the first times Flying A made use of the Palace Theater in this fashion. Located on State Street (the main thoroughfare) near Canon Perdido Street, the

Palace welcomed Flying A personnel in the mornings and afternoons if they needed to screen their films.[16]

One of the Mutual attractions in 1914 was the *Mutual Weekly*. This newsreel presented the events of the day—or rather, week. Each film distributor used the newsreels as an added attraction for audiences. The Pathe organization became one of the first to develop this tool in 1912. They were quickly followed by Mutual later that same year. The earliest mention of newsreels in the *Morning Press* came in 1912. One advertisement by the Mission Theater in the May 4 issue mentions the *Pathe Weekly*. Before the regular newsreels appeared, films presented isolated news events such as the sinking of the *Maine*, or, as in the Mission Theater's May 2, 1912, program, *The Wreck of the Titanic*.[17]

The change from live vaudeville acts to motion pictures fostered competition among theaters. The Portola Theater in Santa Barbara carried a variety of motion pictures but decided in 1913 to remodel its interior for audience comfort and to create the prestigious appearance needed to compete with four other theaters. The Pathe, Lubin, and Essanay productions appeared in the Portola as General Sales Service releases.[18]

One interesting variation on the theater concept is found in a 1913 advertisement—"Free Motion Picture Show Every Night at the Sign of 'Big Board.'" Under this title within the advertisement was the line "advertising that advertises." All of this was presented by the Sunset Advertising Company on the corner of State and Ortega Streets. With a projector behind the screen (as used in the very early projection days of the 1890s), a projected image appeared on the corner of State and Ortega every evening. Short films may have been used to attract people to the advertisements on slides or film. A 1913 article in the *Morning Press* shed some light on this question. Describing a change of program with an Imp drama, *Hard Cash*, the article presented a listing of several short subjects from independent film studios (Imp, Lux, Eclipse, Great Northern).[19] The "Big Board" was located across from the *Morning Press* office. This practice can be traced to New York City in 1897, when J. Stuart Blackton and Albert E. Smith projected "magic lantern" images from slides on canvas. Blackton and Smith's Commercial Advertising Bureau prospered for a time with the addition of motion pictures to their program.[20]

Santa Barbara's Theaters and Motion Pictures

•

After a year of production in Santa Barbara, things were going well for the American Film Manufacturing Company. The studio had facilities rivaled only by Universal in Hollywood. The city of Santa Barbara favored this new industry in its midst. Theaters expanded their offering of motion pictures and increased in number. Santa Barbara, with its own studio, had a special interest in the success of the new enterprise. City fathers realized that the motion picture industry provided excellent advertising for the community. But decisions were already being made that would adversely affect the future of the Flying A studios.

NOTES

1. See Robert C. Allen, "Motion Picture Exhibition in Manhattan, 1906–1912: Beyond the Nickelodeon," in *Films Before Griffith*, ed. John L. Fell (Los Angeles: University of California Press, 1983) 162–75, for a case study of the transition to moving pictures in vaudeville houses.
2. *Morning Press* (Santa Barbara), 1 July 1908, 5.
3. Ibid.
4. T.A. Church, "San Francisco, Cal. Dates Back to the Year 1894," *Moving Picture World*, 15 July 1916, 399–402. See also Geoffrey Bell, *The Golden Gate and the Silver Screen* (New York: Cornwall Books, 1984), 100–101.
5. *Morning Press* (Santa Barbara), 1 October 1910, 4.
6. *Morning Press* (Santa Barbara), 19 October 1910, 2.
7. *Morning Press* (Santa Barbara), 20 November 1910, 2; 26 November 1910, 2; Church, 399–402. See also Bell, 100–101.
8. *Morning Press* (Santa Barbara), 11 June 1912, 2; 6 September 1912, 2.
9. *Morning Press* (Santa Barbara), 4 May 1912, 2.
10. *Morning Press* (Santa Barbara), 17 August 1912, 1, 5; 3 October 1913, 2.
11. *Morning Press* (Santa Barbara), 3 September 1912, 2; 31 May 1913, 2.
12. *Morning Press* (Santa Barbara), 7 December 1913, 2.
13. *Morning Press* (Santa Barbara), 6 December 1913, 2.
14. Timothy J. Lyons, *The Silent Partner: The History of the American Film Manufacturing Company 1910–1921* (New York: Arno Press, 1974), 77.
15. *Morning Press* (Santa Barbara), 19 April 1913, 2.
16. *Morning Press* (Santa Barbara), 17 September 1912, 2.
17. *Morning Press* (Santa Barbara), 2 May 1912, 2; 4 May 1912, 2.
18. *Morning Press* (Santa Barbara), 15 May 1913, 10; 6 December 1913, 2.
19. *Morning Press* (Santa Barbara), 8 April 1913, 2; 22 May 1913, 6.
20. See Charles Musser, "The American Vitagraph, 1897–1901: Survival and Success in a Competitive Industry," in *Film Before Griffith*, ed. John L. Fell (Los Angeles: University of California Press, 1983), 29–32, for a description of the Commercial Advertising Bureau.

CHAPTER 4

Santa Barbara and the Flying A

WHEREAS THE early nickelodeons attracted the lower-income classes, more artistic feature films became an accepted entertainment for the middle and more literate classes. The development of the feature film (four reels, or forty-eight minutes or more in length) began when Adolph Zukor introduced the *Passion Play* from Europe in his Newark Theater. Europe had produced many earlier feature films, and Zukor saw the potential market for features in the United States. His investment in features struck against the trust because the trust did not believe that American audiences would sit through these "long" films. The trust had always broken up the European features into chapters, to be shown over two or three nights. During the presentation of the *Passion Play* (a forty-five-minute feature film), Zukor's organist in the theater played religious music. Zukor states in his autobiography that he observed reactions of audiences to this film. "Many women viewed the picture with religious awe. Some fell to their knees. I was struck by the moral potentialities of the screen."[1] This changed the image of films from vaudeville-type entertainment for the lower-income classes to "respectable" entertainment for the middle and perhaps the upper classes. Then Zukor partly financed the production of *Queen Elizabeth* in France, starring Sarah Bernhardt. By that time, producers had encouraged established stage performers to appear in feature films of their stage successes, again adding "respectability" to the film theater.[2]

With the cooperation of Broadway producer Daniel Frohman in hiring legitimate stage actors for his film enterprise, Zukor established a film company entitled Famous Players in Famous Plays. For a time he even had Edwin S. Porter (famous for *The Great Train Robbery*) direct some of these feature films. The first legitimate theater to show feature films was Frohman's Lyceum Theater, in July 1912; the film was *Queen Elizabeth*.[3]

Thus the star system developed along with the feature film. Studios attracted notable stage actors to the screen to immortalize their stage successes with very high salaries. On-screen billing of the actors' names became a success as audiences looked for their favorites. Zukor paid the stage actors twice their regular salary for appearing in his films. James Hackett starred in *The Prisoner of Zenda*, the first five-reel feature. Zukor also had such stars as Mrs. Minnie Fiske in *Tess of the D'Urbervilles*, James O'Neill in *The Count of Monte Cristo*, and Lilly Langtry in *His Neighbor's Wife*.[4]

At first, the whole industry was apprehensive about accepting the feature film. The belief that audiences would not sit through films four reels or more in length frustrated directors. D.W. Griffith struggled with his parent studio, Biograph, over the release of his four-reel *Judith of Bethulia* (filmed 1913, released 1914). Company officials felt that the public would not tolerate such a long program; thus, Mutual released only one reel at a time on successive nights. Although Griffith pointed out the success in the United States of such European features as *Quo Vadis* (1912) and *Cabiria* (1913), the American film industry was slow to move in this direction. Finally, Griffith propelled the industry toward features after he produced the mammoth *Birth of a Nation* in 1914, which was released in early 1915. The film appeared at the Potter Theater for seven days in May 1915. Under the original title, *The Clansman*, the motion picture received ample advance advertisement in the *Morning Press*. The success of this twelve-reel production significantly altered the motion picture business's attitude toward longer films. Features became an essential part of the industry, and those studios which remained hesitant to produce the longer films were doomed to failure.[5]

During the second half of 1913, the American Film Manufacturing Company cautiously expanded into the two- and three-reel market

Potter Theatre Week Commencing
W. T. WYATT & CO., Lessee
H. CALLIS, Manager. Mon. Eve., May 26

Matinees Every Day, Commencing Tuesday

SEE

Decisive Battles of the Civil War!
Sherman's March to the Sea!
The Burning of Atlanta!
Lee's Surrender at Appomattox!
What It Cost Mothers, Wives, and Sisters!
The Assassination of President Lincoln!
The Rise of the Ku Klux Klan!
The Coming of the Prince of Peace!

D. W. Griffith's
Marvelous Photographic Spectacle
in Twelve Reels

THE CLANSMAN

--- OR ---

"The Birth of a Nation"
Produced by D. W. GRIFFITH
The World's Foremost Producer
from the Novel by THOMAS DIXON, JR.
Depicting the Organization and Motives of the Famous

Ku Klux Klan

Cost to Produce $500,000
18,000 People, 3,000 Horses

— CAST INCLUDES —

Henry Wathall, Mae Marsh, Miriam Cooper, Josephine Crowell, Spottiswood Aitken, Ralph Lewis, Lillian Gish, Elmer Clifton, Robert Harron, George Seigmann, Walter Long, Mary Alden, Joseph Hennebery, Sam de Grasse, Howard Gaye, Donald Crisp, & Jennie Lee

"Mr. Griffith's representation makes 'Cabiria' and 'Quo Vadis' insignificant by comparison." — *New York World*.

SEATS NOW ON SALE
Evenings — 25c, 50c, 75c. Box Chairs, $1.00
Matinees — 25c, 50c. Box Chairs, 75c
Matinee Daily, 2 p. m., Beginning Tuesday

Potter Theatre advertisement of The Clansman (or Birth of a Nation) in the Morning Press, May 30, 1915.

and occasionally even produced four-reel films. This conservative attitude in the face of the rise of features brought a decline in the fortunes of American Film. In the next couple of years, a major split within the Mutual distributing organization resulted in a limited market for films offered by Mutual, and an exodus of talented directors and actors to Los Angeles brought hardship upon the Flying A studio.

The main source of the American Film Company's conservative attitude toward features was its president, Hutchinson, and his partner, John Freuler (also vice-president of Mutual). But Mutual, headed by Harry Aitken, wanted American Film to expand into the feature market in 1914. Charles Hite, another executive with Mutual and president of Thanhouser films, maintained some peace between the other executives. Tremendous difficulties ensued between the executives of Mutual and American Film after Hite's death in August 1914. Despite pressure from Mutual, American Film remained reluctant to adopt the feature. Aitken and Freuler could not reach an agreement on what was best for both Mutual and American Film Manufacturing in Santa Barbara.[6]

Without entering the feature market, American Film under Hutchinson and Freuler began to make some changes in the variety of films produced by the studio. In early 1914, the company commenced a gradual shift from one-reel westerns to comedies. That trend increased after Hite's death. These light comedies, the "Beauty" films, became available each week to Mutual and the theaters by March 1914. Each week, one western, one "Beauty" film, and two social comedies arrived in the theaters on four different days. Attempting to break away from its original specialty, the western, American Film then began to offer a variety of more sophisticated and complex films.[7]

Finally, in March of 1915, American Film entered the feature market. With its release of *The Quest*, the studio hoped to launch a series of features to be known as the "Master Pictures." But Harry Aitken, president of Mutual, refused to release the picture. He explained that it "was not up to standard."[8] Naturally, Freuler and Hutchinson responded with a challenge to Aitken at the next Mutual board of directors meeting. In the May 1915 meeting, Freuler (who had gained backing from Mutual investors) replaced Aitken. But unfortunately for the future of Mutual, Aitken left the organization with his personally owned film companies. Included among these was Majestic-Reliance, which had controlled the Griffith films since the director's break with Biograph. In addition, Aitken's friends, Adam Kessel and Charles Baumann, withdrew from Mutual with their companies, New York Motion Picture and Keystone. This meant that Mutual lost the films of the popular directors: D.W. Griffith, Thomas Ince, and Mack Sennett. These directors then created the Triangle Film Corporation to produce and distribute their motion pictures. This tremendous blow to the fortunes of Mutual affected the future of the American Film Manufacturing Company.[9]

During 1915–1916, Santa Barbara theaters reflected the film industry's struggle with the transition to feature productions. The 1914 Supreme Court decision dismantling the Edison Trust helped independent motion picture companies survive and become stronger. In the next few years, the merger of a motion picture company with the distribution arm of the business ensured a studio's survival. Santa Barbara audiences witnessed this change when they looked at the *Morning Press* theater billings.

The advertisement for the Mission Theater in March of 1916 described its Triangle program. The films of Sennett, Ince, and Griffith had been presented at the Mission since December of 1915. In addition, the theater occasionally carried Mutual releases. The Argus Theater carried the new Jesse Lasky-Paramount releases in 1916, and many of the popular Cecil B. De Mille films. Metro and William Fox releases added to the program. By 1916, the Potter Theater accepted the motion picture as legitimate entertainment for its audiences. Following the success of *The Clansman*, Griffith's *Intolerance* appeared at the Potter Theater in December of the following year. Sarah Bernhardt starred in a motion picture of the stage play *Jeanne Dore* in March of 1916. The Potter selected only the more prestigious films of the period on special road-show engagements. The Portola Theater showed Essanay, Selig, and other smaller studio productions. The *Selig Tribune Weekly* appeared on a regular basis. By the end of 1916, the Palace Theater presented Metro and William Fox films. Notices of Flying A releases were less frequent, for although the Flying A produced many films from 1915 to 1916, competition was exceedingly strong. Chaplin, Griffith, De Mille, and others commanded the box office. The *Morning Press* advertisements reflected this fact.[10]

Following the major Mutual controversy, Freuler led Mutual, and

Hutchinson remained president of American Film. During this period, the Flying A enjoyed the success of its first serial. Serials in theaters had begun a few years earlier in Europe as a series of films with the same leading characters in continuing adventures. Developed in Great Britain in 1909 with the *Lieut. Rose, R.N.* series of one-reelers, these serials increased in popularity in France, with the *Nat Pinkerton, Detective* series and a western serial, *Arizona Bill*. The United States entered the field with Edison's *What Happened to Mary?* in 1912. Many of the early serials were shown in conjunction with newspaper and magazine serials. Newspapers realized circulation increased when their current serial adventure coincided with the screen selection that week. *The Ladies World* collaborated with Edison in *What Happened to Mary?* The success of this six-reel series made a star of Mary Fuller. This serial was followed by the six-episode *Who Will Marry Mary?*, which included a separate, inclusive adventure in each chapter. Not until Selig's *Adventures of Kathlyn*, in 1913, with a *Chicago Tribune* collaboration, did a series with the ending, "continued next week," begin. Each two-reel episode starring Kathlyn Williams brought climactic adventure to the screen. This started the notion of serials with cliff-hanger endings. Edison followed with a third Mary Fuller serial, *Dolly of the Dailies*.[11]

The Mutual corporation reacted with a Reliance company production of *Our Mutual Girl*. These one-reel episodes continued for fifty-two weeks. Norma Phillips starred in the story of a young girl's rise from obscurity to high society. The Palace Theater of Santa Barbara, being a Mutual theater, carried this serial. One advertisement in the June 5, 1914, issue of the *Morning Press* listed the twelfth episode of *Our Mutual Girl* as one of the evening's attractions, along with other Reliance, Thanhouser, and American Beauty films.[12]

Meanwhile, the *Chicago Tribune* formed an association with the Thanhouser company to produce a new serial, *The Million Dollar Mystery*. This serial received Mutual distribution, as Thanhouser was one of their contracted studios. The stars were Florence La Badie, Marguerite Snow, and James Cruze. Because of the success of this twenty-three-episode serial, the *Chicago Tribune* developed another adventure, the thirty-two-episode thriller *The Diamond from the Sky*, produced by the American Film Manufacturing Company in Santa Barbara. Released in May of 1915, it starred William Russell, Charlotte Burton, Irving Cummings, and Lottie Pickford (sister of Mary Pickford). One of the directors was William Dean Tanner (later known as William Desmond Taylor).[13] Released during the same period as *The Clansman's* arrival in Santa Barbara, it produced record lines in front of the Palace Theater. The *Morning Press* credited the unusually high turnout for the third chapter to the "interest taken in the special stunts."[14]

During the production of *The Diamond from the Sky*, the *Morning Press* produced numerous articles in addition to the "'Flying A' Notes" column to help Santa Barbara residents keep track of the film's progress. Roy McCardell, a successful photoplay writer, joined the Flying A company for this production. He had won $10,000 in the screenplay contest held by the *Chicago Tribune*. This announcement was followed two days later with a full biography of McCardell in the "'Flying A' Notes."[15]

Filming for the first episodes occurred in a variety of locations. The Gibraltar Dam and "the tunnel" (referring to Mission Tunnel, which carried water from the dam through the Santa Ynez Mountains to Santa Barbara) appeared in early chapters. Some scenes were filmed on the Santa Ynez side of the mountains above Santa Barbara. Articles in the *Morning Press* described the stunts performed by the company:

> The action calls for an unusual departure from the train, and this was staged on one of the regular north bound trains yesterday morning. Al Thompson got on the train at the depot, while cameras had been placed at a point just south of Hope Ranch, where there is a high embankment. Thompson made his own arrangements for leaving the train, and not even the passengers knew what he was up to. But of a sudden they were startled to see him rush to the front of the coach and open the door of the vestibule and about all they saw was a human figure flying through space. It was about sixty feet to the bottom, and so well was everything done that the thrilling action got right before the camera. Thompson struck in brush and stones, but received only some minor scratches, and thought it was lots of fun.[16]

"'Flying A' Notes" for the following month reported upcoming stunts for *The Diamond from the Sky* for any interested readers. There was to be a fistfight held at the studio in connection with the serial. Besides this notice of filming to come, the *Morning Press* described the recent filming of an underwater octopus fight and another fall from a moving train. With so much advance publicity over the months, the Santa Barbara theatergoer anticipated each chapter of the serial at the Palace.[17]

The Diamond from the Sky's first chapter appeared just before the departure of Charles Hite from Mutual and the disruption of the Mutual organization. The serial helped meet the need to broaden the program of the American Film Company. Along with the "Beauty" films, the serials offered variety and helped deemphasize the westerns. At the same time, serials met with the acceptance of the producers at American Film who resisted a strong effort in the feature field. A year later, American Film released another popular serial. Entitled *The Secret of the Submarine*, it dealt with war preparedness. As with the first serial, *The Secret of the Submarine* garnered advance publicity with the *Morning Press* announcement of filming in various locations.[18]

By the fall of 1915, American Film and Universal were the two best equipped studios on the West Coast. By this time, American Film was releasing one "Mutual Masterpiece" per month, one "Beauty" reel, a two-reel Flying A western, and a one- or two-reel drama. In addition, three-reel "American Star Features" (later the "Clipper Star Features") appeared, and Flying A westerns became known as "Mustang" films. Having different units working on the American Film lot meant that the studio released eight or nine films a week.[19]

American presented sensational or controversial films. In 1913 the French stage play about the dangers of syphilis, *Damaged Goods*, was released. The husband-and-wife acting team of Richard and Joan Bennett starred in this film directed by Thomas Ricketts. Usually, the conservative management of American Film prevented the studio from exploring such explosive topics. But more sophisticated audiences looked for topics of moral controversy instead of the Victorian "black and white" values of earlier motion pictures. But this did not alter the reception of *Damaged Goods*. Many reviewers found the film "a tremendous shocker," and it became one of the most controversial films of the pre-World War I era of motion pictures.[20]

Flying A had a controversial actress in *Purity*, a seven-reel feature, in 1916. Filmed by Roy Overbaugh, this motion picture presented

(Left to right:) Harvey Clark, Clarence Kolb, and Max Dill, October 1916.

Audrey Munson in a starring role. She had gained earlier notoriety with her nude modeling for a statue displayed at the Panama–Pacific Exposition in San Francisco and for the statue of Evangeline at the Longfellow monument at Cambridge.[21] Now she was "living with her mother [and was] the most famous model in this country."[22] Motion picture studios, including Flying A, contracted Munson for work in numerous films. The *Morning Press* announced a Mutual release of one of her Thanhouser films with the comment that "when something of real class comes and one tells the truth about it, it is liable to be discounted."[23] In the Thanhouser film *Inspiration*, she portrayed a young country girl who became a model for an artist. "The picture must prove a real inspiration to those who love art in its higher and finer form…[even though]…there actually was talk at one time of censoring it [the picture]."[24]

By May 1916, the American Film Company included twelve units. The company announced hopes of having a payroll within a year of over one million dollars. President Hutchinson stated that there would be "more features [with] big stars." Kolb and Dill (a vaudeville team), Mary Miles Minter (fresh from her success with Metro), and Richard Bennett would be starting work under American Film contracts. Hutchinson announced,

> We have been giving very close study to this situation [having star features] and will give the people what they want. This may be considered a change of policy, but changes will have to take place in the film industry and only the companies will survive who are quick to adjust themselves to them.... It must not be understood that we will let the rest of the program suffer. Our one, two and three-reel stuff is very popular.... Money will be no object when stories and directors are desired.[25]

After Mutual's loss of Aitken and his film companies, American became the major company in the Mutual organization. Unfortunately, Mutual could not survive with just one strong company.[26]

American's weekly payroll increased from $1,000 in 1912 to $19,000 in 1916.[27] There were "eighteen permanent directors, 75–100 players, 150–500 extras, and 200–300 technical personnel."[28] The year 1916 was the period of peak production for the American Film Manufacturing Company in Santa Barbara with 242 motion pictures.[29]

Mary Miles Minter, accompanied by her mother, Charlotte Shelby, became a very popular figure in Santa Barbara. The *Morning Press* announced her arrival in May 1916 as one of the "Prominent Beauties of [the] Profession.... She is of the winsome type, the kind that picture patrons never tire of."[30]

Mary Miles Minter made twenty-six motion pictures for American Film, working under the direction of James Kirkwood, Lloyd Ingraham, Henry King, and Edward Sloman. Minter remained with Flying A until joining Paramount's subsidiary, Realart, in 1919 as a replacement for their departing star, Mary Pickford. Regarded as a "Fairy Princess" by *Picture Play Magazine*, Minter's career collapsed after rumors of her involvement in the William Desmond Taylor murder of 1922. But her popularity had already begun to decline as Paramount's substitute for Mary Pickford.[31]

Mary Miles Minter.

Ike Bonilla recalls Miss Minter from the time he worked as a bellhop at the Arlington Hotel. Minter's mother, Charlotte Shelby, kept her daughter on a tight rein. Whenever Minter was missing, her mother searched everywhere for her. Bonilla remembers:

> She'd be missing...and she'd [Minter's mother] come tearing down there, and she'd say, "Is my daughter registered here?... Oh, you'd know if she's here or not, I'm sure she's here, I know she's here." And then she'd pull out some name. It was Cassidy or Kerrigan or someone. "Was he here?" We learned a lot.... She'd get so mad at us, she'd say, "I know that she's here!"[32]

Mutual Corporation was undergoing some drastic adjustment with the departure of Charles Hite and his companies from the distributing organization. Freuler, the newly chosen president of Mutual, responded with the signing of Charles Chaplin for a one-year contract. Chaplin was given his own studio, called Lone Star, in Hollywood to fulfill the terms of the contract. But in 1917, Chaplin left Mutual for First National and further fame. With his departure, the fortunes of Mutual went into gradual decline because of the lack of a strong product. The star system was becoming a bankable factor in the industry, and Mutual's studios had very few of the popular stars. The fortunes of Mutual had a strong impact on the destiny of American Film.[33]

During 1914, while American Film experienced real growth in the industry, another film company established itself in Santa Barbara. This locally owned company became known as the Santa Barbara Motion Picture Company. Its wealthy investors, primarily from the neighboring community of Montecito, hoped to profit financially from the new motion picture industry. Under the general managership of Dr. Elmer Boeseke, the studio at 1425 Chapala Street was

> purely a local one in every particular, having no connection in any way with any other company now in the field for the production of motion picture films: Santa Barbara capital only being interested in this corporation, therefore the control will be vested entirely in its local officers and directors.[34]

Boeseke announced that "no pains or expense [would] be spared to secure the best theatrical talent as well as to ensure the finest photographic productions known in the history of the motion picture business."[35] The *Morning Press* announced that this new enterprise would bring in money to the community, as the payroll would be nearly $1,200 weekly. The announcement further stated: "It will afford permanent employment for quite a large number of people who will make their homes here, and will naturally spend most of their earnings in Santa Barbara."[36]

The four-reel productions of the Santa Barbara Motion Picture Company would

> afford greater publicity to Santa Barbara as a home of good pictures and its beautiful scenic settings will be prominently brought before the general public throughout the country, and it is the intention of this company to produce only the highest class of feature pictures.[37]

By 1914, it clearly was in the interest of the Santa Barbara city fathers to utilize the motion picture companies to advertise the attributes of their city.

The *Morning Press* considered the Santa Barbara Motion Picture Company important enough to begin a column similar to the "'Flying A' Notes" on July 2. Entitled "Ess Bee Film," it reported the daily events surrounding the production of motion pictures at the new studio. The first report in the "Ess Bee Film" column stated:

> The first use made of the studio of the Santa Barbara Motion Picture Company at 1425 Chapala Street, was of an interior setting of a scene in the four-reel picture now being taken by the company. It was found that the available light in the studio was particularly fine and the possibilities of getting pictures was far beyond the ordinary from a photographic standpoint...the company [will] do expert photographic work on the premises and with the aid of a printing machine of the very latest type just received from Los Angeles, they will be able to make their own positive films at the studio without having to send the negative films out to some larger city to be

printed.... Everything is now moving very smoothly at the plant and it is a scene of activity every morning when the company is ready to start out for taking pictures as though they had been located there for several months instead of the short time that the company has been getting under way for business.[38]

Production commenced on June 29 with personnel from different companies. The American Film Company lost their chief cameraman, Roy Overbaugh, to this new enterprise. The *Morning Press* revealed this information in the "'Flying A' Notes" column of June 7.

Roy Overbaugh resigned as cameraman with the American to accept a similar position with the Santa Barbara Motion Picture Company. Mr. Overbaugh has been with American about three years and came here with the original company from La Mesa two years ago.[39]

Another director, Lorimer Johnston, left Flying A for the Santa Barbara Motion Picture Company. Other departing actors and actresses included Jack Nelson, who had previously worked in both the Thanhouser and Selig studios; Jane Scott, a local actress; and Caroline Frances Cooke, who had appeared in Charles Frohman's stage productions.[40]

The Santa Barbara Motion Picture Company should have been slated for success with people such as Overbaugh, judging from his subsequent success. Overbaugh later joined Kalem in Santa Monica for a few months upon the collapse of the Santa Barbara Motion Picture Company. Then he resumed work with Flying A until May 1916. After his marriage, he joined the Fine Arts company in New York City and worked with his old friend Allan Dwan. In 1920, he was the cameraman on his most famous film, *Dr. Jekyll and Mr. Hyde*, with John Barrymore, under the direction of John Robertson, for the Famous Players-Lasky company.[41]

In spite of the good press and the outstanding personnel, the company had a poor beginning and never met with success. Roy Overbaugh stated that Boeseke "was the prime mover and several Montecito families poured in money into the stock company."[42] Perhaps the interest of the Montecito families was one reason for the short life of the studio. "Montecito stockholders had the feeling their wives and daughters should naturally have the leading roles in all productions."[43] Their first picture, The *Envoy Extraordinary*, was "quite a pretentious costume picture."[44] Although it was an "extravagant production," Mrs. Overbaugh stated in an interview that she did not "think the pictures were very good."[45]

Despite the demise of the Santa Barbara Motion Picture Company, its creation indicates the opportunity that investors believed existed in the motion picture business in Santa Barbara. Another company encouraged investment from Santa Barbara citizens in June of 1914. The *Morning Press* article of June 2 reported that the "Major Film Manufacturing company…will have motion picture studios in Santa Barbara." The president of the company, Benjamin Moffat, stated, "'We intend to produce only the very best of multiple reel special features and talking pictures.'" The article explained that the company was in negotiations for directors and actors. "'When the Major company is ready to commence production, we expect to have one of the most capable operation forces ever gathered together under one company.'"[46]

The Major Film Manufacturing company had its headquarters in Los Angeles. In the June 5 *Morning Press* came this announcement:

> With more than a score of prospective investors in watchful attendance, the first demonstration of the Major Film Manufacturing Company's talking motion pictures…was given here.… Many Santa Barbara people are interested in the company.… Tonight another demonstration was given for the benefit of a large crowd which was anxious to see the talking motion pictures in actual operation.[47]

Talking pictures were in the experimental stage in 1914. The idea was that synchronization was possible between a recording and the film. Problems remained throughout these years because the synchronization was not always perfect, but some investors believed record and film synchronization was an improvement over purely silent film. The *Morning Press* reported this information in 1914:

> Not one of those who saw the demonstration failed to be convinced that perfect synchronism or spontaneous action

existed between film and record. As soon as the laboratories for producing records can be completed a complete [sic] one reel motion picture, with speaking parts for the actors reproduced on records, will be produced.[48]

Apparently, the enormous expense involved in supplying theaters with the equipment and effectively keeping the recording and film in synchronization prevented Major Film from achieving success.

Another company showed interest in Santa Barbara as a location for filming. In January of 1917, Charles Frohman, the theater impresario who had lamented the loss of his audience to the lower-priced vaudeville-motion picture houses a few years earlier, invested in a venture to produce films of famous stage plays. He created a new company, the Empire All-Star Corporation, to collaborate with Mutual in producing these films. The *Morning Press* reported a Los Angeles meeting between Frohman and John Freuler regarding this arrangement, and it speculated that since Freuler was financially interested in the American Film studio, it seemed likely that some of the films would be produced at the Santa Barbara studio. But, as in the case of the Major Film company, apparently nothing came of this idea.[49]

Even the comedian Chester Conklin considered Santa Barbara as a possible site for a studio. In 1920, Conklin contacted the Santa Barbara Chamber of Commerce for information regarding the area's benefits for film production. The *Morning Press* quoted Conklin as stating, "Los Angeles was becoming overcrowded."[50] But Santa Barbara did not become a permanent home for the famous comedian.

Santa Barbara's theaters underwent change in the 1917–1921 period. The Potter Theater accepted more motion pictures from Universal and Selig. The Argus, a "Marcus Loew house," connecting it with the theater chain from Los Angeles, presented Lasky-Paramount motion pictures. In October of 1917, the Argus closed its doors, to be reopened within a month under the title of Lloyd's Theater, with Mutual productions. The Portola Theater continued to show Mutual and vaudeville programs. The Mission remained with Pathe and Paramount releases. The Palace Theater included the new Triangle films in its program.[51]

To better understand the Flying A studio's decline, it should be noted that during this period of transition, local Santa Barbara theaters were changing film distributors and switching to films produced by studios that controlled their own distribution. The change was indicative of a nationwide trend. Any studio still distributing through the Mutual organization, such as Flying A, discovered an increasingly limited market for its products.

By 1918, the Potter occasionally presented Fox motion pictures and Thomas Ince films. This theater apparently retained several independent studio productions from Southern California. Lloyd's showed Metro releases as well as Mutual films, whereas the Palace dropped Mutual films and showed the First National films of Chaplin during this period. The Portola also stopped showing Mutual pictures and began screening Universal films. The Mission continued with Paramount releases such as the De Mille films.[52]

Despite its difficulties with distribution, the Flying A studio was making good motion pictures during the 1917–1919 period. One director, Edward Sloman, produced a motion picture starring Mary Miles Minter, *The Ghost of Rosie Taylor*, that Kelvin Brownlow describes as "convincing, fast-moving, and expertly directed.... The cutting [was] rhythmic, swift, and informative."[54] The exterior photography was "beautiful." Although this five-reeler was described by Sloman in an interview as "not one of [his] mightiest efforts,"[49] the quality of the film is highly regarded by motion picture historians.

The film's star, Mary Miles Minter, was one of Flying A's biggest attractions and even rivaled Mary Pickford for a time as the sweet young actress on the screen. With an effective director, her performances were quite good. If the direction was weak, her acting suffered. Yet Edward Sloman did not care for her at all. He comments in retrospect:

> Mary Miles Minter was quite young then—sixteen—and very beautiful. Without a doubt, she was the best-looking youngster I ever saw, and the lousiest actress.... I did several pictures with her, but was never happy with any of them.[55]

By 1917, Flying A productions were barely mentioned in the pages of the *Morning Press*. Flying A did not have the major stars or the quality features demanded by audiences at that time. Most theaters in Santa Barbara presented features from major studios in Los Angeles. The theater pages of the *Morning Press* were dominated by extensive reviews, story synopses, and information on some of the leading actors. No longer were there lists of the different short films in a Mutual theater. No longer were Flying A productions listed with Thanhouser or another Mutual studio. By the middle of 1917, the *Morning Press* discontinued the "'Flying A' Notes" column.[56]

Once the United States entered World War I, in April of 1917, the *Morning Press* coverage of events in Europe and Washington, D.C., dominated its pages. The theater section of the newspaper reduced its coverage of the motion pictures at the local theaters over the next couple of years.

By January 1919, only two or three theaters advertised in the *Morning Press*. These included Lloyd's and a new theater, the Strand, which presented Fox productions. During 1920, the three major theaters advertising in the *Morning Press* were the Mission (with Pathe and Fox motion pictures), the Palace, and another new theater, the California. During these years and up through 1921, the Potter presented live entertainment and the Mission continued with motion pictures. The California carried United Artist releases of Chaplin, Griffith, Pickford, and Fairbanks films. On March 21, 1921, the California advertised Chaplin's first big feature, *The Kid*. The Palace reopened on March 3 after remodeling its interior.[57]

There was an ominous advertisement in the January 1921 *Morning Press*: "For Rent—Studio Location—centrally located in business district; well lighted; excellent condition; janitor service."[58] Could this have been the Flying A studios, or perhaps the old Santa Barbara Motion Picture Company site? Nevertheless, indications of the decline of the American Film Company were apparent in the pages of the *Morning Press*. Notices announcing their motion pictures appeared less frequently as the theaters engaged films from the major studios and distributors of Southern California. The discontinuation of the "'Flying A' Notes" column meant less information about Flying A in the newspaper.

Much of the problem with the fortune of Flying A stemmed from the difficulties within Mutual. By 1918, Mutual had lost its major stars, except for those on the American Film Company roster. Many of the stars and directors traveled to Los Angeles between their various films for Flying A. Before long, they realized that the Los Angeles and Hollywood studios offered the best prospects for serious actors and directors (as well as for support personnel). Flying A and studios in San Francisco lost out in the competition for talented professionals. Some major personalities, such as Mary Pickford and Charlie Chaplin, left for greener pastures with other organizations. Pickford and Chaplin later formed their own company—United Artists. Mutual had already suffered a serious blow with the departure of Harry Aitken's group of studios in 1915, thereby losing talented directors Thomas Ince and Mack Sennett.[59]

By 1917, American Film was the major producer of films within the Mutual organization. The serials and selected five-reel features kept the organization strong. Many films appeared in 1916, but there was a precipitous drop in 1917. In 1918, Mutual slipped into indebtedness, and John Freuler relinquished control of Mutual but remained the head of American Film. During 1919–1920, the studio released through a state's rights basis. But this situation merely postponed the inevitable. Finally, without the theaters to release its films, the Flying A studio ceased production on July 7, 1920.[60]

In March 1921, the *Morning Press* reported "rumors that the Flying A might reopen: [The] known facts indicate that the report is correct."[61] The investment by the American Film Company of Chicago and the fact that their equipment had been "kept in thorough order ready for instant use with the assembling of actors and cameramen"[62] led reporters to assume there was some basis to the rumor. Unfortunately, the days of an active studio residing in Santa Barbara had come to an end the previous year. However, the struggling business offices in Chicago continued to exist into 1922. The Flying A's last star, Margarita Fischer, sued the company in February of 1922 for back pay from the last month of her contract in February 1920.[63]

In February 1922, the *Morning Press* reported the use of the back lot of the Flying A studio for a riding stable by a J.J. Gethin, of the Mason Riding Academy. But included was this statement: "It is expected the new film companies soon to begin action next door will take advantage of the...stock near at hand."[64] Marion Davies, actress and longtime companion to William Randolph Hearst, recalled working on a 1923 Goldwyn-Cosmopolitan production in a Santa Barbara "studio which had been empty for years [and] was being fixed up."[65] It appears companies such as the Hearst and Samuel Goldwyn enterprise used the studio facilities during the 1920s. Eventually the buildings and equipment were sold, and although the site was not used for a permanent new studio, nearby locations used by Flying A retained their appeal.

At the beginning of 1925, Roscoe "Fatty" Arbuckle directed Al St. John in *The Iron Mule* at Zaca Lake in the Santa Ynez Valley. In this short comedy spoof of director John Ford's western epic, *The Iron Horse,* Arbuckle used the Pacific Railway tracks near Los Olivos. For his "Iron Mule," Arbuckle even borrowed a small nineteenth-century locomotive from his close friend and fellow comedian, Buster Keaton. This charming little film (which is available for viewing today) illustrated an interest by companies in the Flying A's old motion picture locations. Any notion of a new motion picture studio in Santa Barbara ended when, a couple of months later, the June 1925 earthquake devastated the city. But Santa Barbara recovered, and with her new Spanish-style architecture, this picturesque community beckoned as a possible location for filming future motion pictures.

NOTES

1. Zukor and Kramer, 57.
2. Zukor and Kramer, 56–62.
3. Zukor and Kramer, 66–70.
4. Zukor and Kramer, 75–91.
5. Gerald Mast, *A Short History of the Movies*, 4th ed. rev. (New York: Macmillan, 1986), 61–63; *Morning Press* (Santa Barbara), 27 May 1915, 2; 30 May 1915, 2; 4 June 1915, 2.
6. Timothy J. Lyons, *The Silent Partner: The History of the American Film Manufacturing Company 1910–1921* (New York: Arno Press, 1974), 77–78.
7. Lyons, 79.
8. Terry Ramsaye, *A Million and One Nights: A History of the Motion Picture Through 1925* (New York: Simon & Schuster, 1926; reprint, New York: Touchstone Books, 1986), 718.
9. Ramsaye, 717–18.
10. *Morning Press* (Santa Barbara), 5 May 1915, 2; 3 March 1916, 6; 7 November 1916, 6; 9 December 1916, 6.
11. Anthony Slide, *Early American Cinema* (New York: A.S. Barnes, 1970), 157–59.
12. Slide, 159–60; *Morning Press* (Santa Barbara), 5 June 1914, 2.
13. Slide, 160.
14. *Morning Press* (Santa Barbara), 2 June 1915, 2.
15. *Morning Press* (Santa Barbara), 4 May 1915, 2; 6 May 1915, 2.
16. *Morning Press* (Santa Barbara), 15 May 1915, 2.
17. *Morning Press* (Santa Barbara), 6 June 1915, 3.
18. Lyons, 82–84.
19. Lyons, 80, 84–85.
20. *Moving Picture World*, 2 October 1915, 90; quoted in Kevin Brownlow, *Behind the Mask of Innocence*. (New York: Alfred A. Knopf, 1990), 58
21. Lyons, 85–86.
22. *Morning Press* (Santa Barbara), 25 May 1916, 6.
23. Ibid.
24. Ibid.
25. *Morning Press* (Santa Barbara), 14 May 1916, 8.
26. Ibid.
27. Lyons, 86, 88; *Morning Press* (Santa Barbara), 12 August 1914, 2.
28. Lyons, 88.

29. Lyons, 87.
30. *Morning Press* (Santa Barbara), 19 May 1916, 6.
31. Charles Gatchell, "Concerning a Fairy Princess," *Picture Play Magazine*, March 1920, 31–33, 103; Amos Aydelott, "Mary Miles Minter," *Films in Review* 27 (October 1969): 473–95, 501.
32. Isaac (Ike) Bonilla, interview by author, 15 July 1988, Santa Barbara.
33. Lyons, 88.
34. *Morning Press* (Santa Barbara), 7 June 1914, 8.
35. Ibid.
36. Ibid.
37. *Morning Press* (Santa Barbara), 14 June 1914, 3.
38. *Morning Press* (Santa Barbara), 2 July, 1914, 2.
39. *Morning Press* (Santa Barbara), 7 June 1914, 2.
40. *Morning Press* (Santa Barbara), 14 June 1914, 3.
41. *Morning Press* (Santa Barbara), 20 May 1916, 5; Robert Birchard, "Roy Overbaugh, A.S.C." *American Cinematographer* 65 (May 1984): 34–38.
42. Roy Overbaugh, "Movie Capital," interview, *Santa Barbara News–Press*, 10 August 1958, 11(E).
43. Ibid.
44. Mr. and Mrs. Roy Overbaugh, "In the Days of the Flying A," interview by W. Edwin Gledhill (*Noticias*, 17 March 1954), *Santa Barbara Historical Society* (Fall 1976): 15.
45. Ibid.
46. *Morning Press* (Santa Barbara), 2 June 1914, 3.
47. *Morning Press* (Santa Barbara), 5 June 1914, 2.
48. Ibid.
49. *Morning Press* (Santa Barbara), 17 January 1917, 10; 25 January 1917, 3.
50. *Morning Press* (Santa Barbara), 18 June 1920, 8.
51. *Morning Press* (Santa Barbara), 5 January 1917, 6; 6 January 1917, 6; 4 October 1917, 6; 5 October 1917, 6; 8 October 1917, 6; 13 November 1917, 6; 5 March 1918, 6; 31 March 1918, 6; various other issues of the *Morning Press* during this period.
52. Ibid.
53. Kevin Brownlow, *The Parade's Gone By...* (New York: Alfred A. Knopf, 1968; reprint, Los Angeles: University of California Press, 1975), 156, 158.
54. Brownlow, 161.
55. Ibid.
56. *Morning Press* (Santa Barbara), various issues during the 1917–1918 period.
57. *Morning Press* (Santa Barbara), 13 June 1919, 6; 15 June 1919, 6; 5 January 1921, 6; 3 March 1921, 2.
58. *Morning Press* (Santa Barbara), 8 January 1921, 6.
59. Lyons, 89–94.
60. *Morning Press* (Santa Barbara), 10 March 1921, 1(II); Lyons, 89–94.

61. *Morning Press* (Santa Barbara), 10 March 1921, 1(II).
62. Ibid.
63. Morning Press (Santa Barbara), 2 February 1922, 7.
64. Morning Press (Santa Barbara), 3 February 1922, 3(II).
65. Marion Davies, *The Times We Had* (New York: Bobbs-Merrill Company, 1975; reprint ed. New York: Ballantine Books, 1977), 41–42

✣ CHAPTER V ✣

Conclusion

THE AMERICAN FILM Manufacturing Company failed because of a lack of distribution. Originally formed as a product of the film exchanges, it produced westerns in direct competition with Essanay. The American Film motion pictures supplemented the other Mutual pictures, including films from Keystone, Thanhouser, Majestic, and Lux, among others. The Palace Theater exemplified a Mutual house, showing motion pictures from all the Mutual studios. The Mutual organization received a severe blow, from which it never recovered, when it lost the studios controlled by Harry Aitken. American Film, as only part of the Mutual roster, could not produce the entire Mutual program.

The American Film Company pioneered western location filming, along with the Essanay studios of "Broncho Billy" Anderson. Arriving in Santa Barbara in 1912, Flying A provided strong competition with its varied roster of westerns, social comedies, and serials. However, the distance from Los Angeles became a problem for personnel. Many of the actors and actresses wanted to be close to other studios for employment between Flying A films. Directors and actors needed proximity to the primary activity of the industry. By 1916, Los Angeles had become the focus of the industry. People in American Film gradually accepted contracts with the up-and-coming studios in Los Angeles.

The conservative management of American Film kept the company from being on the cutting edge of changes in the industry. Audiences during World War I became more sophisticated, and tastes changed from the "black and white" morality of westerns to issues of the day. Certainly, Flying A did get into the market with a film on syphilis (*Damaged Goods*) and *Purity*, but these were the exception. American Film suffered from a conservative ownership's reluctance to accept features until it was required for survival. American was not the place to be for a director desiring to be at the forefront of the industry in 1917. The feature film, the sophistication of audiences, the end to distributing organizations with a roster of studios, and the beginning of the star system had a profound impact on the conservative Flying A.

Understanding the history of the Flying A and how it fits in Santa Barbara history is certainly interesting. But what makes the story important is viewing the role Santa Barbara played as a community caught up in the latest "rage" of the times—the motion picture. Its reputation as a picturesque community was enhanced throughout the period as Santa Barbara answered the call for scenic locations for a studio, extras for those many crowd scenes, and a newspaper ready to report it all. And finally, Santa Barbara reflected the manner in which many small communities responded to the new novelty of film with a shift from vaudeville to motion picture entertainment in theaters. As seen in the following appendix, Flying A will be remembered for years as a fascinating industry that brought some exciting memories for the town and its citizens.

✣ APPENDIX A ✣

Memories of the American Film Company

INDIVIDUALS IN THE Santa Barbara area have memories of the days of the Flying A studio. What follows are the transcripts of pertinent portions of interviews with people involved with the American Film Manufacturing Company in Santa Barbara: Loiz Huyck, Mrs. Robert Phelan, Arleigh Adams, and Isaac Bonilla. All four played "extra" roles in the films. Loiz Huyck also had a job as an assistant make-up person in charge of "touching up" for further filming in the course of the day. All provide a personal insight into those early days of motion pictures.

Loiz Huyck

Loiz Huyck, interviewed on August 9, 1987, recalled her youth in Santa Barbara and her involvement with the American Film Manufacturing Company. She worked as an assistant make-up girl on the set after school each day. She occasionally received roles as an extra in ballroom and beach scenes.

Her memories include the child actress on the lot, Baby Helen (Helene Rossen, who worked for Flying A in 1916), and the actress Vivian Rich (who worked there from 1913 to 1916). In addition, Mrs. Huyck recalled the leading man, Jack Kerrigan (1910 to 1913), and the attitudes of the cowboy extras toward Kerrigan.[1]

Huyck: Vivian was a very lovely-looking girl. She knew how to act; she was a good actress and all, but she didn't know anybody, and she didn't like many of the people she had to work with. And her mother didn't approve of any of them. Her mother was a squatty little person. She was a splendid seamstress. She sewed continuously to cut down on her [expenses] because in those days the actors and actresses had to furnish their clothes and...Vivian Rich's mother was at the sewing machine continuously, and if she went anywhere, excepting to her church, she carried a basket of sewing with her and everybody could tell what she was making.... She was just wonderful.... Vivian asked [my] mother if I could come over after school. Instead of coming directly home and sitting around alone, I would come up to the studio and help her. Vivian and I got along beautifully. I was about ten years younger than she was, but I was just full of wanting to help somebody. So I got up there and she taught me how to put make-up on her. She had some [?] things to do and she was trying new make-up. They do that all the time at the studio, and the thing is that I learned to put on a foundation, and I learned to put on eyebrows so they looked natural, and I learned to put on eye shadow and where...Vivian would have me...remake her make-up. She taught me to retouch because during

Vivian Rich in front of the Arlington Hotel.

those days—some of them were very hot and all—her make-up would melt a little. The thing is, I would learn how to repair make-up. That was very difficult because after make-up's been on, if you try to retouch it, it's very violent. The colors are very brilliant immediately, somehow.

So then after I learned to put make-up on, they wanted [me] to at least help the make-up man, who was putting on make-up on these wild wahoos who were playing cowboys. All of them wanted to look as handsome as the leading man at the studio, who was...Warren Kerrigan...Jack, I mean—was Jack Kerrigan...And the thing is, it wasn't in the woods for them to be handsome at what they were doing. And I tried to talk them out of it so that they would accept themselves as they really were. And some of them were really rugged-looking men and made money later in "B" [movies], rugged mountaineers and things like that, character parts. I said, "If you can do a good character part, until you get handsome, that's fine, but don't try to be handsome first thing off. And don't make me try to make you handsome." I said, "I can exaggerate your coloring and do you a little bit of good, but I am not going to try to make you so handsome."

Lawton: They put a lot of make-up on the actors in those days....

Huyck: Oh, they did.

Lawton: But wasn't that because of the way it was filmed?

Huyck: That was the only way in which the film would take them in the old days.... Well, they were good actors, I guess. But not what I'd call, well, [laughter].... But the thing is, I'd come in after school and these boys would be working all day, and I'd be having to retouch them. The retouching was to me harder than putting a full make-up on. But I'd have to do that in between their takes—whatever they'd be doing. They'd be sitting around, waiting for...to do a cowboy something, and I'd have to retouch them and all. It was interesting, but those cowboys made my life miserable. They always wanted to look like Warren [Jack] Kerrigan. Or they always wanted to look like somebody else. I said, "Aren't you satisfied being you?" And the thing is, that it wasn't much of a job. I mean, I wasn't paid outrageously or anything of the sort, but it was the first money I ever earned....

Mrs. Huyck went on to describe her childhood in Chicago. Her mother had visions of her daughter being a great pianist, so the young Loiz was forced to take piano lessons. She remembers being an expert artist. She made beautiful paper dolls. She sewed many small outfits for the dolls. She also recalled that her two aunts and her cousin, all single, dressed extravagantly for social occasions.

Mrs. Huyck arrived in Santa Barbara during 1904, at the age of twelve. She went to the boys' school instead of the private girls' school because she was not happy with the girls' school and left it for her brother's school. She lived in the guest room of the head teacher and his wife. The classes were more academic than at the girls' school. She was raised by her grandmother and great-grandmother because her mother was an opera singer who remained in Chicago for a time after Mrs. Huyck moved to Santa Barbara.

Lawton: Could you tell me a little bit about what you remember of the studio, how it worked, and what it was like there at that time?

Huyck: Well, it was a very "homey" sort of a place as far as I'm concerned. Excepting that I didn't know until I'd been there for a few days that we didn't walk into Vivian's area, and she didn't walk into Warren Kerrigan's area without a phone call in advance. And you had to come when they were willing to let you come.... You see, I was too young to realize the implications of this business.... I knew what had to be done before they went before the cameras. That's practically all I knew...

...This little tiny girl [Baby Helen] was the bossiest little thing—three years old...and Warren Kerrigan or one of the men would have to get down on their hands and knees and carry her around, let her ride them horseback. Then she'd do something. She was the biggest hit on radio and television and things like that because nobody knew what a bossy little thing she was. She was always sweet, loving, and kind and all. And she said, "You know, I really took advantage of those nice people."

Lawton: This is Vivian Rich's daughter?

Huyck: No, no, nothing to do with Vivian. No, this is the little girl that was called Baby Helen. And she said, "People keep coming and asking me, what did I do and all? And how did I put my make-up on? I didn't do that. I didn't dress myself. I didn't put my make-up on. Everybody had to wait on me hand and foot." And she said, "These little tiny children get pretty far when they're waited on hand and foot like that."

Then when [Baby Helen] moved down south, her aunt and uncle were the ones who were taking care of her. For some reason or other, her mother [and father weren't] available, and they were taking care of her and they were banking the money. And they disappeared into thin air, leaving this child homeless.

Lawton: That happened a lot with some of the child actors.

Huyck: Yes, and that's why the court.... This was before there was court action. And she was stranded with no money, no idea of business, and then she became a very smart little businesswoman. Because when Thomas Ince [the film director] died, her backing died [1924]. He was backing her as a good actress. She was a smart little thing and all. And the thing is, she was good when she wanted to be. But she loved to be bossy.

I saw her many times. Let's see, I was sixteen, or eighteen, when she was three. So, I'm ninety-five now...so she must be eighty.

Lawton: She is still a friend of yours?

Huyck: Well, yes, but the thing is, we're not what you'd call intimate friends. She went into three or four things that she did. But she liked dry cleaning. And she found she could manage that, and she ran and owned a dry cleaning place when she finally retired...from business. She's still very lively and very bright.

Lawton: Did you ever walk out to where they were filming and watch it all?

Huyck: No, the only thing I ever joined in was, I was an extra several times in ballroom scenes. I dressed beautifully and I had loads of clothes. And we wore long dresses with trains when we were sub-debs.

Lawton: Does that mean a very young girl?

Huyck: Well, that means you're under eighteen. And the thing is, I guess I worked as an extra on a party and beach scenes and things like that.... And I'd always drag in a few of my friends so I'd have somebody to talk to because I wasn't going to stand there and be dumb. And the thing is, it was fun and it paid a little money, but not too much. I mean, mine was only a couple of hours. Because I'd have to go home and change my clothes and get ready for supper.... I had to be home by six. My mother for about seven to eight months didn't know I was doing this. She thought I was doing extra work at the college—no, the high school. I was on the basketball team and I was on the tennis team. And I could always...[say] I was playing tennis.... [I'd] come home with a racket when I'd be telling her I'd been playing tennis. And I'd come home properly wet, with a wet bathing suit, if I said I was in swimming or something. I felt dreadfully about it for a while. But after a while, the fact that I was doing make-up became a natural thing to me and I told her. "Now, look, I've been studying this with the Flying A up there. Vivian has taught me how to put make-up on for the camera. I've learned and [I'm] being paid a little bit of money for it." And she said, "That is all right, that was a job, but don't neglect your studies." So that was it.

Mrs. Huyck stated that she got married in 1917 and did not make a career of her make-up. She learned to do clay work at a pottery school and married the instructor. Her large home above Valerio Street contained a room large enough for frequent dances.

Lawton: From where you lived and down to where the studio was, I imagine there was a lot of country and then the road would go down to the heart of the central part of Santa Barbara.

Huyck: Yes, when the Flying A bought that area above Mission, it was considered pretty far out. And we had mule-drawn streetcars. One went to the Mission, the other went to Oak Park. I don't know if it was called Oak Park. But anyhow, it was the park area toward the Cottage Hospital—beyond the Cottage Hospital....

The social life of the Flying A people was pretty stringent. A lot of the people in Montecito didn't approve [of] the movies. They snubbed them and all. But they secretly envied them, I think. But the thing is, Edgerly Court was practically the center of the activities—social activities of the movie people because Edgerly Court was built at the same time they were building the Flying A. And they had a great big ballroom. I don't know whether it's still there or not. But the thing is, the moving picture people practically took that area over. They either lived at the Upham [Hotel] or Edgerly Court with an apartment of their own. If they were married, they could live at Edgerly Court because they'd have someone to do their cooking. There was no restaurant attached to the Edgerly Court area. And most of the moving picture people went over to the Upham to eat or to the restaurant.... But there were kitchenettes attached to every little apartment. They were quite livable. And...these youngsters who knew any of them at all would go there [at night]. Every night there was a dance. It was our only medium of social exchange for the youngsters because most of us were pretty well supervised, and we didn't go out on [the] beach...or wahooing around in the automobiles. My brother...took me over there, and he usually stayed and took me back to the house. But we had a lot of fun. It was quite unusual for us.

Mrs. Robert Phelan

Leontine Birabent, born in 1893, married Robert Phelan, a cameraman with the Flying A studio. She performed as an extra in various roles. Her abilities as a swimmer earned her a place in many of the scenes requiring a young lady with aquatic skills. In a telephone interview on March 30, 1988, Mrs. Phelan related her impressions of the Flying A company.

Mrs. Phelan stated that the American Film Manufacturing Company arrived in Santa Barbara because they "were looking for a variety of scenery." Upon using the Montecito estates as locations for filming, the studio frequently presented the owners with still photographs of their homes and the studio company.

They were a "very friendly lot of people." They "had their families there. They accepted us as we accepted them." There was "never any problem." The studio "employed local people." The city welcomed them because of the financial benefits. For one thing, "they needed rentals and there was purchasing of all kinds." They lived in two or three cottages and one main boarding house.

Mrs. Phelan's husband began as a film editor and graduated to cameraman. He was a cameraman with the Flying A studio from 1914 to 1915. He photographed *Purity*, the motion picture which starred Audrey Munson, who had posed for statuary displayed at the Panama–Pacific Exposition in San Francisco.

At first the film negatives were sent out of Santa Barbara to Chicago for developing (or perhaps to La Mesa). Later Mrs. Phelan explained that the studio processed its film in Santa Barbara. Many times the editing was done from the negative.

Arleigh Adams

Arleigh Adams took residence in the only remaining building (except for the garage) of the Flying A studio. His family purchased the dressing room and waiting room (green room) which the actors used before their directions and scenes of the day. Mr. Adams recalled the studios from his perspective as a child extra actor who lived across the street from Flying A. Before the interview, he showed the author the dressing rooms of the long building, which are bedrooms today. He used the original kitchen and bathrooms. Outside the building is an overhanging trellis extending the length of the building within a long courtyard. This trellis remains from the days of Flying A. Mr. Adams was interviewed on July 16, 1987, within the waiting room, which included a fireplace. This room served as his living room.

Lawton: You moved here in 1906 or 1907?
Adams: Well, we were still on Brinkerhoff Avenue. We didn't come up...[until later]. This was a cow pasture when I was a kid here. I'm eighty-three now. I was about seven years old when...[I moved here in] 1911. Well, that's when [sic] they came in here and built. They came from Chicago and went to Orange County, and then they came up here on location. This was more wild country than it was down there. So they moved up here, and they were down here on State Street and rented a place there. This [Mission and State streets] was all a vacant lot. They came up here and bought this whole lot, the whole square block and half the next block.
Lawton: Was this an ostrich farm at one time?
Adams: No, that was where they were before. That was an ostrich farm. They raised ostriches to have plumes for the ladies' hats. Well, it was on the corner of Islay and State street[s]. And then they came up here and built this. And seventeen companies [film units] ended up here. There was a lot of activity going on.
Lawton: You didn't live across the street at that point?
Adams: We lived here before they built this because I used to play on

this lot and there was cattle in here.

Lawton: What did it look like? Were they pretty busy?

Adams: Yes, they were busy, and they'd come over here. Some of the stars—the Morrison brothers—they lived over here in apartments. They were cowboys.

They'd [the studio] come over and ask us to take pictures of our house, and all of us kids were around then and we'd get to watch....

Lawton: Do you remember coming inside this building [the green room]?

Adams: Yes, and they'd call you and I'd come over here. And sometimes they weren't ready for you. And sometimes you'd sit here all day long...and I got three dollars a day. That was a lot of money when you were seven years old. And if you did bit parts you got five. So, then I got into a scene where I was supposed to play marbles out there in the street. A steamroller came up the street and was supposed to run over me. They run it up to me and I put my clothes out here and I was supposed to be flattened out....

Then they had *The Diamond from the Sky*. It [the diamond] was supposed to come out of the sky and hit a house up here and blow the whole thing up at the end of State Street and Constance Avenue. That was all vacant then.... Well, the whole town turned out to see that thing. There wasn't anything left. It really blew it. That was a big show.

The film was probably *House of a Thousand Scandals*, a four-reeler released on September 23, 1915. Made during the same period as *The Diamond from the Sky*, the film was given extensive coverage in the *Morning Press*. The article also mirrored the excitement recalled by Mr. Adams over seventy years later:

> About 2000 people witnessed the taking of this sensational scene for *The House of a Thousand Scandals*.... Spectators commenced to appear early, and there were about 200 automobiles lined up on Constance Place.... During the noon hour the charges of powder had been placed. This consisted of

150 pounds of giant powder and 50 pounds of black.... The thing went off with a bang that could have been heard for miles, and at the same time the mess shot upward. This spread out in a fanlike formation and was very black, just the thing for a picture. It is estimated that the outline of the heaving mass went to a height of 500 feet and then it settled.... Everyone agreed that it was a first-class explosion.[2]

Adams: And then I got into another one called *Love and Green Apples* [*Green Apples*, released August 31, 1915.[3]] I was supposed to be a kid who always ate these green apples and get sick in bed. Then camera, action—they started up.... But the kids start clowning around...I started grinning and blew it. He kicked me out of the bed and looked for a "sicker" kid. That ended my career in the movies right there.

But my dad played in them. And my sister played in them.... *The Lost Chord* [*Trail of the Lost Chord*, released in two reels on November 17, 1913.[4]] I forgot the leading man, and he played that chord and she [Adams' sister] was supposed to be a peasant girl. And she'd stand there by him when he was playing the lost chord. And then my dad played in them. My mother didn't play. My brother played in them, too—Indian pictures. I have a picture of that. They stamped the Flying A on the pictures [photographs].

Lawton: Did you do this in the summer or after school?

Adams: Saturdays—there wasn't any school, so I could work right across the street. They took a lot of pictures of the houses [in the neighborhood].

Lawton: You've probably been all around the complex. What did it look like? What kind of things did it have?

Adams: [As Mr. Adams viewed his photographs] It took this whole block and half of the next block.... [They] had a small glass studio right off of Mission Street. Then later on they built a big one.... [Notice] the angle they built this thing. They didn't have klieg lights, so they used the morning sun and afternoon sun.... [There was] a carpenter's shop...and there was a garage.... [There were a lot] of cars. There was a Winton, Cadillac, La Salle, and Model T Ford. And [there were a lot]

Automobiles of Flying A.

    of chauffeurs for the stars....
Lawton: You moved into the building in what year?
Adams: We [my wife and I] came in here in '32.... My dad bought the building.
Lawton: What was the building used for before your family bought it?
Adams: It just laid idle. When the [Flying A] moved out, it just sat here. It was the Depression time, and they couldn't sell anything around here.
Lawton: The rest of the buildings were still here on the lot then?
Adams: Yes, and they were just gradually sold off.... Someone wants to buy this place [where he lives now]. The garage was just sold about a month ago for $139,000. I could have bought that for $2,400 when we bought this. But we just didn't have any money....
Lawton: Where did they do the filming that they didn't do here [at the studio]?
Adams: They went all over up in the hills here. The Botanical Gardens and off Mountain Drive, because this was rugged country. Nothing was built up, you know.
Lawton: This was wide-open land?

Adams: Yes, you could buy this land for a dollar an acre. You can't believe that now....

Lawton: Where did the people who worked here live?

Adams: All over. Mary Miles Minter was down on Garden Street, and Edgerly Court...that was a hotel court...a lot of them lived there. And the Morrison brothers—they lived right across the street. The Morrison brothers were cowboys.

Lawton: Were there any big hotels downtown?

Adams: The Potter Hotel. That was down at the beach. That was a beautiful building. It got all burned up [1921]. And the Arlington—where the Fox-Arlington Theater is [today].... Now that burned down, too, in 1922, I think it was...[it was 1909]. And then the '25 earthquake came and knocked it down and that finished it....

Lawton: Did your family get to know any of the people who worked for the studios?

Adams: Well, no, not too much. There was Kolb and Dill [a vaudeville team that made pictures in 1916 for Flying A]. They had a Stanley Steamer out here, and boy, when they started that thing up, it shook the house....

There was a restaurant on the corner right across the street [from the studio].

Lawton: When were most of the homes built in this neighborhood, in this part of State Street?

Adams: ...They were nice stucco buildings. They had a nursery here. They had half of a block that is named American Avenue [today].... This is where they run their horses. They had a nursery for the kids....

Isaac Bonilla

Isaac "Ike" Bonilla, born in 1903, resided in Santa Barbara most of his life. In an interview March 31, 1988, he recalled his contact with the American Film Manufacturing Company. He helped as a stable boy in the corrals and later worked at Pacific Produce and at the Arlington Hotel as a young boy. At this store, he came into contact with the Flying A personnel. As he recalled the Flying A studio, he turned the pages of a scrapbook containing numerous photographs of Flying A and a list of Charlotte Burton's films. Most of this scrapbook was given to him by the actress and her mother.

Bonilla: [They had] a big glass stage, enclosed glass stage which was supposedly at that time the largest in the United States....
[One of the actresses] Charlotte Burton...was a Santa Barbara girl. The daughter of Benjamin Burton, who was quite famous...in the early years of Santa Barbara....

The productions were made in Santa Barbara from 1912 to 1916...were movies in which Charlotte had a role in or was a supporting bit. In addition to these pictures, she had a lead in three serials: *The Diamonds in the Sky*, *The Three of Hearts*, and *The Secret of the Submarine*.

Mr. Bonilla explained that the studios sent articles to the local newspapers to obtain adequate coverage of their work.

He worked at the Arlington Hotel as a bellhop and delivered cablegrams. This job allowed him to observe many of the personnel from the studio. Mary Miles Minter had a home in Santa Barbara, where she lived with her mother. Many of the actors and actresses gathered at the Fior de Italia, a beer garden on Presidio Avenue. Many of the people working for the studio lived at Edgerly Court on Chapala and Sola Streets.

Filming occurred throughout Santa Barbara. Mr. Bonilla mentioned Veronica Springs in the Las Positas area. This was used for the westerns; it had the advantage of being not too far from the studio. In addition, Hope Ranch and Mirago Springs Ranch were used as locations.

Mr. Bonilla noticed the declining fortunes of the studio when he worked at Pacific Produce and realized that many of the actors could not pay their bills during the last couple of years of the studio's operation. He stated that the director Henry King had his credit cut off in 1919.

The Argus Theater presented short programs of motion pictures, and the Potter Theater, along with the Lobero Theater (the Santa Barbara Opera House), were used at times to show films.

A second interview followed on July 15, 1988.

Lawton: Did you know Charlotte Burton?

Bonilla: Yes, as a messenger boy, I worked for her mother, Mrs. Burton. And she [Charlotte] lived at home at that time…at 601 De La Vina. And Mrs. Burton was a very high-grade type of a person.… Her father was a newspaper man and an All-American track man and in the Hall of Fame at the University of Texas.… Charlotte Burton, of the people that they used [in the films] out of Santa Barbara as actors and actresses, she was the lead one in the most films. She [was] in *Diamond in the Sky* that had…many episodes.…

Lawton: Did you work for Diehl's Grocery Store or Pacific Produce?

Bonilla: Yes, I worked for Pacific Produce Company, which was a…mixture between a San Francisco-style raw product produce and groceries from San Francisco for the people.… We handled a full line of groceries.… Diehl's was an all-round grocery store—a high-class grocery store that went the limit for you. You call them up from New Haven and say, "I'm going to be in Santa Barbara December the first, and I want a six-room house and three baths and a corral for my horses, and he would have it when [you] got here. They furnished everything. They were two different kinds of grocery stores. I worked for the lower class, but did the most business.…

Lawton: Did you work for the studio at all? Were you a stable boy?

Bonilla: Yes, we did everything—whatever. Now, for instance, our family lived on Modoc Road. Now, that was a natural background for the Flying A. There was a working cattle ranch there. There were real cowboys—three of them.… They did their branding and everything right next door to where the

Memories of the American Film Company

Postcard view of State Street, showing Diehl Grocery Company.

La Cumbre Junior High is now. They had background right there. Now, the rolling hills back of Oak Park; the Presidio, from San Francisco—the San Francisco Presidio used to come down once a year and camp there; put up their tents and camp there. They ended up with a "sham" battle...make-believe...empty their shells, shooting up and down those hills.... They entertained the city of Santa Barbara, the actors did.

Lawton: They did a lot of things for the city—getting involved....

Bonilla: Getting involved, yes, well...the Flying A didn't have to go very far—it wasn't two or three blocks—to get the people that they wanted. Now, it's true they had...these big-shots there, but when it came down to the guys that played the part, they didn't have to go too far.

Lawton: Did you remember coming in contact with any famous people from the studio? Actors or actresses?

Bonilla: Well, when you say come into contact with them....

Lawton: Delivering messages?

Bonilla: Cummings. Cummings—now, Cummings was the guy who would hire us kids. [Irving Cummings, actor and cowboy with Flying A in 1915.⁵] He'd keep track of us on how many hours we'd work. He'd pay us off or give us a lunch. Ice cream. That was our favorite. Now, when I say these kids, in my neighborhood within a quarter of a mile in a circle you could round up...from nine to twelve kids overall...you could use as barefoot boys, never wore a shoe until they had to....

Mr. Johannson had a working cattle ranch at the far end of La Cumbre Junior High.... They had working cowboys.... The Flying A would come over, get these cowboys, and use them and the background. The Flying A backgrounds [were] natural deep arroyos.... So we didn't have to make believe there at all. They had the ground....

There was a railroad going through Santa Barbara, and you could walk faster than it went. They had a steep grade going from the depot up till you got to Veronica Springs before it stretched out, and then [they] could open her up. So they did a lot of their railroad scenes with what they had right there.

Lawton: Was that the Southern Pacific?
Bonilla: Yes, Southern Pacific.
Lawton: When I was doing research in the newspapers...the *Morning Press*, I noticed the theater advertisements for the Potter Theater and the Argus Theater, Palace Theater....
Bonilla: The Potter Theater was live, with plays and George White's *Scandals*....
Lawton: That was the better theater.
Bonilla: Yes, all the better shows showed there and...it was a live show theater.
Lawton: There was one called the Santa Barbara Opera House. Is that where the Lobero is now?
Bonilla: Well, yes, that's right....
Lawton: The Lobero closed earlier, back in the late 1800s.
Bonilla: Yes.
Lawton: Then it reopened again as the Lobero.
Bonilla: The Opera House was the one in Chinatown. It was well built. Jose Lobero came down from San Luis Obispo, and he brought my grandfather with him....
Lawton: You could tell when the studio wasn't doing so well. You told me once that the director Henry King—they had to cut off his credit?

Bonilla: Yes, they had to go out and sell their stuff....
...Across from the old Los Angeles creamery on Ortega between Chapala and State Street, they used to show live pictures on a screen right out in the open.

Lawton: I ran across advertisements in the paper. They called it the "Big Board"....

Bonilla: Yes, behind the white house in there.... [Looking at a photograph of the automobiles of the Flying A] Now, these are the cars...Win-Sixes, Locomobiles, Pierce-Arrows. There's a big Win-Six here....

Lawton: I also ran across the fact that they started another studio, the Santa Barbara Motion Picture studio. It was started by...Boeseke?

Bonilla: He had...the first polo field. And then one of his boys, the younger boy, he was a giant. He was seven foot tall; big Elmer they called him. Elmer Boeseke became a star in the United States later on in polo.

Lawton: ...Elmer...was acting in some of the Flying A—Elmer and his pony...

Bonilla: Oh, yes, he was a natural anyway....

Lawton: You mentioned that you were working in front of the Arlington—

Bonilla: Yes....

Lawton: Mary Miles Minter had a room there....

Bonilla: No, her mother always refused it....

Lawton: So, she didn't have a room there.

Bonilla: Well, I don't know if she did or not. We were never allowed to tell her [Minter's mother]. "Go up and see the room clerk; he's the one that knows her. I'm just the bellhop." But my station was in the basement and they had a stand-up bar.... Oh, she [Minter's mother] would come down there. She'd be missing, Mary Miles Minter would be missing—she didn't show up...and she'd come tearing down there, and she'd say, "Is my daughter registered here?".... "Oh, you'd know if she's here or not, I'm sure she's here, I know she's here." And then she'd pull out some name. It was Cassidy or Kerrigan or someone. "Was he here?" We learned a lot.... She'd get so mad at us; she'd say "I know that she's here!"...

Lawton: Flying A had its own baseball team?

Bonilla: Yes. [Mr. Bonilla showed the author photographs of filming in Summerland. The picture of the main street of Summerland was Highway 101.]

Now, there was a Ferguson girl who played a double for Mary Miles Minter. She had long blond curls and she was pretty. And she'd stand in for a lot of her....

[Another photograph of filming on the stage of the Potter Theater. Mr. Bonilla thought it was a scene from *Trilby*.]

...And Irving Cummings, they were awful good to us kids.

Lawton: Were they actors, or actors and cowboys—both?

Bonilla: Actors. [But] the Morrison boys came right off the cattle ranch. And as far as I know, most of these guys were bow-legged. They were real cowboys. And they could perform and they could go up to Lompoc. Fourth of July they had a big rodeo up there. Several of them participated. They were working cowboys.

...The champion steer bulldozer of the world...1912. He was there.... He was in the Flying A. "Shorty" McNall; now, he was a real working cowboy. He settled here in Santa Barbara.... "Buddy" Baer, Max Baer's younger brother...[was also there as a cowboy].

Kolb and Dill. Now, they were stage people, but they also did movies. But they were just a two-man act.... Now, we had a chorus; the Orange Grove Review, I think it was called.... We had a chorus at the Mission Theater that was there all the time.... And then, we brought in the Mack Sennett Bathing Beauties.... They showed the latest thing in...bathing suits....

At this point, Mrs. Bonilla mentioned that she was in one of the Flying A productions. She was one of many kids in school used as background in one of the films. But she never saw it, "because we weren't allowed to see movies."

Lawton: I noticed in the paper they were very careful in the articles they wrote to try to convince everybody that these people were okay....

Mrs. Bonilla: Yes.

Bonilla: [looking at a picture] Now, this is the open air bar and dining room and everything. Joe Sequoia's [on] East De la Guerra Street and Presidio Avenue...Fior d'Italia, we called it. And they took a lot of pictures here because it was a typical Italian...he had a big mustache; he was a typical Italian.

[Looking at a still photograph from one of the many "Calamity" Jane films made by the Flying A] This is "Calamity" Jane....

Lawton: They made a series of those?

Bonilla: Yes. [Viewing another photograph] Now, they had a Model T Ford that they hauled...extras in.... They hauled in about twenty people. [Another photograph] Now, this is my uncle. He played the part of a detective...Mariano Bonilla. He ended up as a...maitre d' at the Biltmore Hotel.... [Looking at a picture of the California Theater].... Now, this is not Santa Barbara; I think this is San Francisco. The California [in Santa Barbara] did not have a marquee like that. It had a little sign....

Lawton: I ran across [in the *Morning Press*] different theaters like the Palace Theater.

Bonilla: The Palace was movies.

Lawton: Movies only.

Bonilla: The Argus was movies. Strictly. The Mission was movies and vaudeville. The Portola was movies and vaudeville. And they were in big competition with each other over who had the best act. The Potter was strictly stage....

[The back lot of the studio was rugged country.] They used that a lot. That and Las Positas [an area nearby]. The bull pasture, we called it, the bull pasture. Well, in fact, this whole range—this property here—the whole thing from Haley Street to Carrillo Street...

Lawton: Was the California the old Argus Theater?

Bonilla: No. The Argus was a little tiny thing. [Referring to locations] Now, they did a lot of waterfront [scenes]. Big sailing ships would come in here with freight. We had, about every other month, a big square-rigger in here....

Lawton: Santa Barbara seemed to be strongly behind the motion picture company. From reading the paper, I get the impression they were for it.... Was this true?

Bonilla: Well, for a while. They built the Edgerly Court. They went heavily in debt for that. That was strictly for the housing of the Flying A.... We only had one big apartment house...and the Edgerly was a real modern...it catered to the needs. And then they got heavily into debt....

[Referring to Robert Phelan, the cameraman with Flying A] Now, he was a freelance photographer. I mean, he did everything in movies. He had his own [equipment].... I think he came with the Flying A, but he ended up freelancing....

Henry King [the director] and some of those I got to know later on—they were always friendly.

NOTES

1. Morning Press (Santa Barbara), 12 December 1913, 2.
2. *Morning Press* (Santa Barbara), 8 May 1915, 2.
3. Timothy J. Lyons, *The Silent Partner: The History of the American Film Manufacturing Company 1910–1921* (New York: Arno Press, 1974), 235.
4. Ibid.
5. Lyons, 254.

BIBLIOGRAPHY

BOOKS

Allen, Robert C. "Motion Picture Exhibition in Manhattan, 1906–1912: Beyond the Nickelodeon." In *Films Before Griffith*, ed. John L. Fell, 162–75. Los Angeles: University of California Press, 1983.

Bell, Geoffrey. *The Golden Gate and the Silver Screen.* New York: Cornwall Books, 1984.

Bowser, Eileen. *The Transformation of Cinema: 1907–1915.* New York: Scribners, 1990; reprint, Los Angeles: University of California Press, 1994.

Brownlow, Kevin. *Behind the Mask of Innocence: Sex, Violence, Prejudice, Crime: Films of Social Conscience in the Silent Era.* New York: Alfred A. Knopf, 1990.

————. *The Parade's Gone By...* New York: Alfred A. Knopf, 1968; reprint, Los Angeles: University of California Press, 1975.

————. *The War, the West, and the Wilderness.* New York: Alfred A. Knopf, 1979.

Davies, Marion. *The Times We Had.* New York: Bobbs-Merril Company, 1975. Reprint ed., New York: Ballantine Books, 1977.

Gish, Lillian, and Ann Pinchot. *The Movies, Mr. Griffith, and Me.* 1969. Reprint ed., Englewood Cliffs, N.J.: Prentice–Hall, 1975.

Griffith, Mrs. D.W. (Linda Arvidson). *When the Movies Were Young.* 1925. Reprint ed., New York: Benjamin Blom, 1968.

Koszarski, Richard. *An Evening's Entertainment: The Age of the Silent Feature Picture, 1915–1928.* New York: Scribners, 1990; reprint, Los Angeles: University of California Press, 1994.

Lyons, Timothy J. *The Silent Partner: The History of the American Film Manufacturing Company 1910–1921.* New York: Arno Press, 1974.

Mast, Gerald. *A Short History of the Movies.* 4th ed. rev. New York: Macmillan, 1986.

Musser, Charles. "The American Vitagraph, 1897–1901: Survival and Success in a Competitive Industry." In *Films Before Griffith*, ed.

John L. Fell, 29–32. Los Angeles: University of California Press, 1983.

———. *The Emergence of Cinema: The American Screen to 1907*. New York: Scribners, 1990; reprint, Los Angeles: University of California Press, 1994.

North, Joseph H. *The Early Development of the Motion Picture (1889–1909)*. New York: Arno Press, 1973.

Ramsaye, Terry. *A Million and One Nights: A History of the Motion Picture Through 1925*. New York: Simon & Schuster, 1926; reprint, New York: Touchstone Books, 1986.

Roske, Ralph J. *Everyman's Eden: A History of California*. New York: Macmillan, 1968.

Slide, Anthony. *Aspects of American Film History Prior to 1920*. Metuchen, N.J.: Scarecrow Press, 1978.

———. *Early American Cinema*. New York: A.S. Barnes, 1970.

Southworth, John R. *Santa Barbara and Montecito: Past and Present*. Santa Barbara: Orena Studios, 1920.

Starr, Kevin. *Inventing the Dream: California Through the Progressive Era*. New York: Oxford University Press, 1975.

Tompkins, Walker A. *Santa Barbara Past and Present: An Illustrated History*. Santa Barbara: Tecolote Books, 1975.

Zukor, Adolph, and Dale Kramer. *The Public Is Never Wrong: The Autobiography of Adolph Zukor*. New York: G.P. Putnam's Sons, 1953.

PERIODICALS

Aydelott, Amos. "Mary Miles Minter." *Films in Review* 27 (October 1969): 473–95, 501.

Birchard, Robert. "Roy Overbaugh, A.S.C." *American Cinematographer* 65 (May 1984): 34–38.

Church, T.A. "San Francisco, Cal. Dates Back to the Year 1894." *Moving Picture World* 15 (July 1916): 399–402.

Cool, Margaret. "The Flying A Days." *Santa Barbara Magazine* 3 (Fall 1977): 26–32, 34.

Gatchell, Charles. "Concerning a Fairy Princess." *Picture Play Magazine*, March 1920, 31–33, 103.

Gidney, C.M. "About Santa Barbara County." *Overland Monthly* 38 (August 1901): 157–72.

Henry, William M. "The Great God Kerrigan." *Photoplay Magazine*, February 1916, 32–36.

"In the Far West." *The Bioscope*, 9 February 1911. Quoted in Anthony Slide. *Early American Cinema*, 66–67. New York: A.S. Barnes, 1970.

Lawton, Stephen R. "The Early Theaters of Santa Barbara." *Noticias*, Summer 1992.

Morning Press (Santa Barbara), 1908–1922.

Moving Picture World, 8 February 1913, 559; 10 July 1915, 248; 15 July 1916, 399–402.

Nielsen, Michael C. "Labor Power and Organization in the Early U.S. Motion Picture Industry." *Film History: An International Journal* 2 (June–July 1988): 121–32.

Overbaugh, Roy. "Movie Capital." Interview. *Santa Barbara News-Press*, 10 August 1958, 10–11(E).

Overbaugh, Mr. and Mrs. Roy. "In the Days of the Flying A." Interview by W. Edwin Gledhill, *Noticias*, 17 March 1954. *Santa Barbara Historical Society* (Fall 1976): 1–2

Packhurst, Donald. "Broncho Billy and Niles, California: A Romance of the Early Movies." *The Pacific Historian* 26 (Winter 1982): 1–22.

Smith, Wallace E. "Santa Paula's Film Days." *The Ventura County Historical Society Quarterly* 16 (Winter 1971).

MISCELLANEOUS

Adams, Arleigh. Interview by author, 16 July 1987, Santa Barbara.

Bonilla, Isaac (Ike). Interviews by author, 31 March and 15 July 1988, Santa Barbara.

Gish, Lillian. Letter to author, 5 March 1988.

Huyck, Loiz. Interview by author, 9 August 1987, Carpinteria.

Phelan, Mrs. Robert. Telephone interview by author, 30 March 1988.

The main gate to the Flying A compound.